MUSTANG 5.0 *Projects*

**PERFORMANCE AND UPGRADE
HOW-TOS FOR 1979-1995 5.0 MUSTANGS**

MARK HOULAHAN

HPBooks

HPBooks
are published by
The Berkley Publishing Group
A member of Penguin Putnam Inc.
375 Hudson Street
New York, New York 10014

First Edition: October 1997

The Penguin Putnam Inc. World Wide Web site address is
http://www.penguinputnam.com

© 1997 Mark Houlahan
10 9 8 7 6 5 4 3

Library of Congress Cataloging-in-Publication Data

Houlahan, Mark.
    Mustang 5.0 projects / Mark Houlahan. -- 1st. ed.
      p.   cm.
    ISBN 1-55788-275-4
    1. Mustang automobile--Performance. I. Title.
    TL215.M8H675 1997
    629.222'2--DC21   97-25716
                                                                                                       CIP

Book Design & Production by Bird Studios
Interior photos by the author unless otherwise noted
Cover photos courtesy *Super Ford* magazine

All rights reserved. No part of this publication may be reproduced, stored in a retrieval system, or transmitted in any form, by any means electronic, mechanical, photocopying, recording or otherwise, without the prior written permission of the publisher.

NOTICE: The information in this book is true and complete to the best of our knowledge. All recommendations on parts and procedures are made without any guarantees on the part of the author or The Berkley Publishing Group. Tampering with, altering, modifying or removing any emissions-control device is a violation of federal law. Author and publisher disclaim all liability incurred in connection with the use of this information.

ACKNOWLEDGMENTS

I would like to thank the following people for, without their help, encouragement, patience, and support, this book would never have been written.

Tom Corcoran, previous Editor of *Mustang Monthly,* an early mentor of mine, and a dear friend; Jerry Pitt and Donald Farr, Editorial Directors at Dobbs Publishing Group, both of whom have had the hard task of keeping me in line from 9 to 5; Larry Dobbs, Founder of Dobbs Publishing and Publisher of *Mustang Monthly* and *Super Ford* magazines, for giving me a chance to prove myself five years ago; Michael Lutfy, Automotive Editorial Director of HPBooks, for giving me the chance to put my foot in the door of book publishing without cutting it off; Steve Turner, Rob Reaser, Pat Hays, and Chuck James, for all their photographic and electronic imaging help; and finally, to my parents, Bill and Joy Houlahan, and my brother, Brent, for supporting me in everything I've ever done. Thank you all from the bottom of my heart.

Last, I would like to say a special thank you to my wife, Dawn, and my children, Kyle and Shelby, for their support, love and understanding. It is to them that I dedicate this book, my first. I truly love you guys.

CONTENTS

INTRODUCTION V

SECTION I: GETTING STARTED 1
   5.0 Buyer's Guide 2
   Garage Setup & Safety 7

SECTION II: ROUTINE MAINTENANCE 9
   Power Windows & Door Locks 10
   Power Timing 15
   HVAC Repair 18
   Replacing Axle Bearings 27
   5.0 Tune-Up 30

SECTION III: ENGINE TECH 39
   Cat-Back Exhaust System 40
   Shorty Headers 44
   MAF Meter Swap 48
   Roller Rocker Arms 52
   Low Temp Thermostat 58
   Throttle Body & EGR 62
   K&N Fuel Injection Performance Kit 66

SECTION IV: FUEL SYSTEM & INDUCTION TECH 71
   Fuel Pressure Regulator 72
   High Volume Fuel Pump 76
   Nitrous Installation 83
   Supercharger Installation 90

SECTION V: TRANSMISSION TECH 97
   Clutch Replacement 98
   Installing a Performance Shifter 105
   Shifter Bushings 108
   Transmission Oil Cooler 111

SECTION VI: SUSPENSION TECH 115
   Performance Springs & Shocks 116
   Strut Braces & Subframe Connectors 122
   Budget Brake Upgrade 127
   Performance Lower Control Arms 137

SECTION VII: INTERIOR TECH 141
   Steering Wheel/Shift Knob Replacement 142
   Body Support Bar 147
   Auxiliary Dash Gauges 151
   Remote Fuel Door Release 157

SECTION VIII: BODY TECH 161
   Installing Fog Lamps & Rewiring the Factory Harness 162
   Swapping Taillights 171
   Installing a Power Antenna 176
   Cobra Grille Insert 180

INTRODUCTION

The Ford Mustang has been an icon of the automotive world since its introduction in 1965. In the first 2 years of production, Ford sold over 1 million Mustangs to people from all walks of life. The Mustang was, and still is, a sporty vehicle with a mystique accessible to everyone. The Mustang could be configured to suit just about any taste, to suit any type of driving. Today, the Mustang carries on the tradition of a low cost, sporty two-door that still seats four.

But since 1979, the Mustang has taken on another image—that of hot rod. In its 5.0 iteration, the Mustang has proven to be a formidable competitor, both on the strip, track and stoplight. Many compare the 5.0 Mustang to the '57 Chevy, another mass-produced vehicle that captivated the hearts and minds of hot rodders the world over. Like the '57, the Mustang has been produced in large numbers, and the aftermarket has produced a vast array of specialty parts and accessories to modify the Mustang for any level of performance, from mild street to competitive racer.

This book is about how to apply some of that technology. It contains 32 how-to "projects" designed to make your Mustang a much better performer for the street. By using the information in this book, you can improve the suspension for better handling; the engine for more efficiency; the interior for comfort. There are tips to keep your Mustang in top running condition, maintenance procedures and upgrades that help you get the most from your ride. Casual enthusiasts and serious street machine performance nuts will find something in the following pages. And, you don't need a 5.0 to benefit. V-6, four-banger Fox owners and V-6 '94-'95 SN-95 Mustang owners will find some of these projects useful.

We've organized the projects into sections to make them easier to locate, and as you flip through each one and study the detailed photos, you'll be surprised at how easy some of these projects are to complete. Some Mustang owners would never attempt to undertake a clutch or supercharger installation, thinking it beyond their mechanical ability. But each of the projects in this book was selected to be completed with the aid of nothing more than a standard set of tools and common sense. Just follow the steps outlined, and you should be okay.

We've kept an eye on the cost as well. With the exception of the supercharger, each of the projects can be completed for less than $500; some for even less than $100.

So grab your tools, this book, and let's head for the garage.

I.
GETTING STARTED

5.0 BUYER'S GUIDE
GARAGE SETUP & SAFETY

5.0 BUYER'S GUIDE

Some of you out there may not even own a late-model Mustang, but you're interested enough in them to spend your hard-earned cash on this book. For others, you bought your first late-model Mustang new, but now you're looking for a second one to build, but you want to buy used and save some money. Whatever your reasons are, we felt a year-by-year guide to late-model Mustangs should be part of this book. Let's face it, consider buying a used late-model Mustang a weekend project in itself. This guide is meant to help you determine what year Mustang you want and what it looks like. We unfortunately just don't have the room to discuss how to buy a used car. Just remember, the majority of 5.0 Mustangs are rode hard and put away wet, so be on your toes at the used car lot.

To begin with, let's define a "late-model" Mustang. A "late model" is defined as any four, six, or eight cylinder Mustang built from the model years 1979 to the present model year, which is 1997 at the time of this book printing. Our Guide will focus on the 5.0 Mustang up until 1995. While the primary focus of this book is the eight cylinder, or "5.0" Mustangs, many of the suspension, body, and maintenance projects are the same for the four- and six-cylinder Mustangs. The late-model Mustang built from 1979 to 1993, or what is known as the FOX Mustang platform, varies little from year to year and most body parts are interchangeable. Thus it is a relatively easy feat to buy a 1982 Mustang and make it look like a 1991 Mustang with nothing more than bolt on changes such as bumpers and mouldings. This holds true also for changing from one trim level to another. Turning a 1989 LX 5.0 into a GT simply takes the factory GT body trim, GT wing, and GT front and rear bumpers.

Let's go over the 5.0 late-model Mustang year by year, detailing the exterior and interior changes so you can acquaint yourself with the Mustang changes over the years and determine which year you should be looking for.

The 1979 5.0 engine was similar to the '82 - '83 models.

1979—The first year for the new Fox platform Mustang, coupe and hatchback body styles only, no convertible, and no GT option. On the outside, the Mustang used 14" wheels (optional metric aluminum "TRX" wheel package), 10" front disc brakes and 9" rear drum brakes, bright window trim, and an egg crate style grille with quad rectangular headlights. The 5.0 engine was only available with a four speed manual transmission. The new corporate 7.5" integral axle was also used on the '79 Mustang. The interior used the "Fairmont" dash board, low mounted door handles, optional console, and the fold down seat in the hatchback is one piece.

1980—Changes for the Mustang included moving the inside door handle to the top of the door panel, and interior and exterior color options. The 5.0 V-8 is dropped and replaced with a 4.2 liter (255 cid) V-8 reportedly for better fuel economy.

1981—The 4.2 V-8 resides again under the hood for

5.0 BUYER'S GUIDE

The '80 sedan 2-door (shown here with fake top) was available with a 5.0 liter V8 with TRX wheel/suspension package.

The '82 GT went for the Trans Am look, with front air dam, blackout trim and rear spoiler.

The '83s sported a new nose, complete with hood scoop. This nose would remain with the Mustang through 1986.

another year on the Mustang. All other engine options and related vehicle designs remain unchanged.

1982—This was the year that touted "The Boss is back" with the reintroduction of the Mustang GT, the first GT badged Mustang since 1969. The GT had twin chrome tailpipes exiting on the driver's side of the rear valance, a 5.0 HO (high output) V-8 engine breathing through a two barrel carburetor and attached to a four speed manual transmission. Electronic ignition with a performance curved distributor, performance gauges, and aluminum wheels rounded out the package.

1983—The GT Mustang continued into 1983, and has remained until this day. The 1983 model year also marked the return of the convertible Mustang, last seen 10 years ago in 1973. The GT would not be available as a convertible until the tail end of 1983, but would be back for a full production run for 1984 and on. The GT lost the old SROD four speed and picked up the Borg/Warner T-5 five speed transmission, which was used in different ratios until 1994, when it was replaced with the T-45. 1983 also marked the first major exterior change as the tail light panel was re-stamped to accept

5.0 BUYER'S GUIDE

The interior from the '80-'86 cars remained pretty much unchanged through all of those model years.

The '86-'88 models came with EFI equipped with speed density meters.

new lenses with separate brake and turn lamps.

1984—For 1984 Ford offered Central Fuel Injection, or CFI on the Mustang 5.0. It was only available, and mandatory, when ordering the Automatic Overdrive transmission, or AOD. The CFI optioned 5.0 was only rated at 165 horsepower, which many people believe was to prevent damage to the non-performance designed AOD. The 1984 Mustang looks identical to the '83 model on the exterior and didn't benefit from any other performance enhancements. 1985 would be the banner year for these items.

1985—The 5.0 HO, as well as the Mustang itself, came off the Dearborn assembly line with some major

The '84-'85 5.0 engine came with a 4V Holley.

changes for 1985. The 5.0 block was revised to accept roller lifters and a roller camshaft was used with a hotter profile than previous years. Tubular exhaust headers instead of cast iron were used, as well as 2 1/4-inch dual exhaust. These changes netted a horsepower value of 210 for this '85 5.0. The Mustang itself was enhanced with gas tuned shocks, variable rate springs, and larger sway bars. The Quad-Shock rear suspension used on the SVO Mustang found its way to the 5.0 Mustang and 15x7 "10 hole" aluminum wheels with Goodyear Eagle "Gatorback" performance tires, which replaced the old TRX metric wheel package, were used for better handling and traction. A short throw shifter was added to the T-5 five speed and the interior was upgraded with newer "performance looking" materials.

1986—1986 marked the first production year of SEFI, or Sequential Electronic Fuel Injection, on the Mustang. The SEFI engine setup used an injector for each cylinder, not just two like the CFI setup, and a tuned long runner intake was used for more torque production. The carburetor would now be a thing of the past on the 5.0. The '86 Mustang lost some horsepower though due to a new high swirl design cylinder head to help emissions, netting a final rating of 200 horsepower. This would be rectified in 1987 when Ford would install the E7TE truck head and gain back the horsepower loss, and then some. A new 8.8" rear end was installed in '86, which could better handle the future power increases of the 5.0 Mustang. The '86 5.0 Mustang still used the quad headlight arrangement from earlier Mustangs, and the

5.0 BUYER'S GUIDE

The most popular model by far are the '87-'93 Mustangs. More aftermarket performance parts are available for these cars than any other model of the Fox chassied generation.

taillights remained unchanged, but the Federally mandated CHMSL, or Center High Mount Stop Lamp (more commonly known as a third brake light) meant a change to the back of the Mustang, where engineers added a luggage rack on convertibles, and modified the rear wing on hatchbacks for the new light.

1987—For some Mustang fans 1987 is considered one of the best years. With the previously mentioned truck heads, a new upper intake manifold, and a larger throttle body, Ford was able to up the ante to 225 horsepower. And if that wasn't enough, Ford did a total redesign of the Mustang's interior and exterior. The outside got new flush mounted aero head lamps with a new nose, the body side mouldings were revised, and the small side glass/louver package was replaced with one large flush side glass. The GT model also received a factory "ground effects" package, round Hella fog lights molded into a GT specific front fascia, and a GT specific rear wing that was raised up off the hatch. The front end geometry was changed with a new front crossmember and larger 11" front disc brakes were added. The interior received an all new dash assembly with a raised "pod" for housing the instruments. New ergonomic power window and lock switches were installed with redesigned door trim panels and a new console and seats.

1988—The 1988 model year Ford simply treaded water with some exterior color change options and the addition of Mass Air Flow metering on California shipped Mustangs. This MAF metering system allowed for direct reading of the incoming air flow, which in turn allowed for more radical modifications to the engine while maintaining good idle quality and emissions.

1989—For 1989 Ford simply added the California based MAF metering system to all 50 states. The 5.0 LX

In '89, the 5.0 was revised with a mass air meter for improved performance. This would be the standard through the '93 models.

5.0 BUYER'S GUIDE

The 1994 model signified the most radical improvement since 1979. Refinement was added to the interior, although power was somewhat compromised. However, the numerous improvements far outweighed these drawbacks. There are plenty of power enhancing options available to the enthusiast.

Sport model was officially announced from Ford, even though you could order a non-GT based 5.0 Mustang for years, and included a few special options.

1990—Ford added some safety to the Mustang in 1990 by adding a driver side air bag (which changed the steering column, making tilt wheel unavailable) and rear three point shoulder/lap belts. Early '90 models were shipped without a center console armrest for weight savings and CAFE enhancements, but public outcry returned the armrest to the Mustang by mid-year. Performance and braking/handling aspects remained unchanged for '90. Unfortunately, Ford never built a 25th Anniversary Mustang in 1990. All that Ford did was add a running horse tri-bar emblem to the dash. This emblem, similar to the fender emblem of a '66 Mustang, stated "25 Years" on the bottom of it. For the first time, since the early '80s, an all black interior could be ordered.

1991—Ford's only exterior change for '91 was the upgrade to 16" wheels and tires. Arguably the nicest set of wheels to come on the Mustang in a long time, the "five stars," as they've come to be known, had a silver finish and the running horse logo in the center cap.

1992—Ford's changes to the '92 Mustang consisted of changing the exterior body side mouldings to body color and the dash gauge pod and steering wheel to match the interior color instead of black as in years past.

1993—For the last year of the Fox platform Mustang Ford changed the pistons from forged aluminum to Hyperutectic pistons. This move was made to quiet the motor, as some owners in the past had complained of piston slap when the engine was cold. Several "Feature" Mustangs were built towards the end of the '93 model year. These Mustangs featured special paint, embroidered floor mats, and painted or chromed factory five star wheels. Production of these cars were limited to just a few thousand each. Ford also introduced the Cobra Mustang in 1993. Available in three colors, the Cobra Mustang featured revised 5.0 engines, 17" wheels, four wheel disc brakes, different ground effects, and almost certain collectability status. The Cobra "R" model debuted toward the end of the year in extreme limited numbers and was meant for racing use only.

1994—The Mustang received a new, more rounded, "'90s" look with a total sheetmetal redesign, while retaining a nearly identical floorpan and firewall arrangement to the '93 and earlier cars. The 5.0 engine was made more compact to fit these engine compartments by lowering the intake and shortening the front end accessory drive and bringing the accessories in closer to the block. Besides the new upper intake and tubular headers, the 5.0 remained relatively unchanged.

1995—The '95 model year is identical to the '94, except at the end of '95 Ford offered a stripped down 5.0 model called the GTS. This model, similar to the older 5.0 LX models, would get you a 5.0 Mustang without the "GT" moniker and with a few less options.

GARAGE SETUP & SAFETY

All through this book there are projects that you will be capable of accomplishing in the average weekend using ordinary shop tools. Tools which include sockets, ratchet wrenches, combination or "boxed end" wrenches, cutting tools, pliers, jacks, jack stands, pry bars, and many other specialty tools. We want you to enjoy and benefit from the projects found between the covers of this book, but not at the cost of personal injury.

Every year thousands of people are seriously injured and even killed in household accidents, a large percentage of which occur in the garage or shop. To prevent injury always practice what I call "The Common Sense Laws of Mechanics." For instance, if you have a nut to loosen, don't even think about using an open ended wrench to break the nut free, use at least a boxed end (enclosed end) wrench or a socket. While 12 point sockets and wrenches are cheaper, six point sets will give you better grip with less slippage. When you do try to break the nut free, use an open palm instead of an enclosed fist around the wrench handle. This way, if the

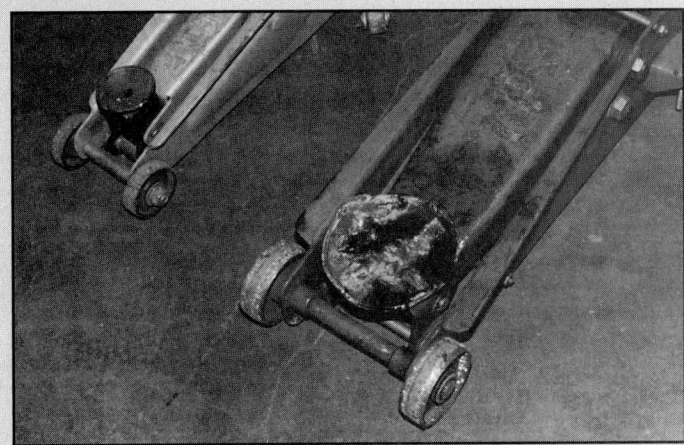

Any tool that has to do with a potentially hazardous operation, such as jacking up your Mustang, should not be skimped upon when purchasing. The two floor jacks you see here are both rated at two tons of lifting power, but which one would you feel safer using? The large six inch head of the red jack is much more apt to keep your Mustang from slipping off the jack than the little three inch head on the silver one. Don't be cheap when it comes to your safety.

nut unexpectedly breaks free, or if the wrench slips, your knuckles won't get busted. I'll show you more of these laws in the accompanying photos.

The work space itself has to be conducive to working on your Mustang too. Don't try to raise your Mustang with a floor jack on an inclined driveway, don't set up jack stands to put your Mustang in the grass on the side of your house, don't paint a part for your Mustang with your wife/girlfriend/buddy's car parked two feet away and the wind blowing towards their car, etc. I know these all sound like "duh, I think I know better than to do that," but if you think back, I am sure there has been a time when we have all worked in less than safe conditions because we were too cheap to buy the right tool, in too much of a hurry to enact the appropriate safety measures, or just plain weren't paying attention. So check out the following safety guidelines and do's and don'ts photos before you go any farther into this book, and enjoy your weekends.

Ensure that all your tools are clean and safe to use. If you have a chrome socket that is cracked, don't use it. A cracked socket is sure to cause a busted knuckle; just get it replaced. Also, if you use air tools, don't use chrome sockets on them. Only use the "impact" type black sockets with air tools.

GARAGE SETUP & SAFETY

While we are on the subjet of jacks and jack stands, don't ever consider this: working under a car with just a jack. A five minute clutch cable adjustment can cripple you or even kill you if the floor jack fails or someone bumps it. It will only take another minute to grab a jack stand and be safe. The two most common types are the "pin" style and ratchet style. The pin jackstands can only adjust to a set number of heights, and the head is usually not designed to fit the center of the vehicle. The ratchet type is cast steel rather than stamped (stronger), and the head allows for centering, such as on a rear axle. Get the stronger ratchet kind if possible.

Safety Guidelines:

• If the job requires a tool you don't own, beg, borrow, buy, or rent the right tool. Specialty tools are just that, special, and regular tools will not accomplish the same job. Regular tools will just damage the fastener, or part, and possibly cause personal injury.

• When working under your Mustang always, and I mean always, have your Mustang in Park for an automatic, or in a forward gear for a manual transmission, the parking brake set, high capacity jackstands locked in place and in the correct location, and tire chocks in use at opposite wheels.

• When working on electrical repairs or modifications, disconnect the battery's negative cable to prevent electrical damage or sparks.

• Don't smoke when working on your Mustang, even if you're not working on part of the fuel system. Cigarette ashes and embers can smolder and start an interior fire, damage the paint, or catch wiring on fire.

• Take breaks often. Every project has a frustrating moment that will get you throwing wrenches into the street. Take five minutes, replenish your body fluids, take a deep breath, and try a new approach to the problem. Frustration is where accidents start and stupid things happen, like dropping a bolt down the engine's intake manifold.

• Always read the instruction manual over several times before attempting the project. This will mentally prepare you for the job at hand, as well as give you time to rent any specialty tools, set up shop time at a machine shop, or get the parts you need before hand. Just think how many times you have pulled a car apart for some work, then found out the local parts house won't have the part you need until the middle of the week.

• Keep a well stocked first aid kit within a few steps of your work area, preferably hanging on the wall of your shop or garage next to a water source. I know it sounds gross and cold hearted, but you don't want to bleed all through the house trying to get to a bandage in the upstairs bathroom.

• Some background music is nice to break the silence when working on your Mustang, but don't have it so loud it annoys the neighbors or people nearby can't hear you call for help if needed in an emergency.

• Speaking of people nearby, try to schedule your work when there will be someone home. If you need emergency help, or even someone to run and get you a part, there is someone there within earshot of your voice.

A clean work area is a must. Having a cluttered area causes problems when attempting to use the workbench. Items get damaged, lost, or dirty. Other tips are to always wear safety goggles while working under the car, especially with a rotary tool such as a die grinder, and keep a fire extinguisher at hand.

II.
ROUTINE MAINTENANCE

POWER WINDOWS & DOOR LOCKS
POWER TIMING
HVAC REPAIR
REPLACING AXLE BEARINGS
5.0 TUNE-UP

POWER WINDOWS & DOOR LOCKS

It is not uncommon to see a late-model Mustang with inoperative power windows or power door locks, or windows that slip and chatter like they're off the track, because of all the options on a Mustang, windows and locks get the most use.

The worst thing you can do to power windows or door locks is install an alarm or remote operation kit with door lock and window interfaces. Most of these kits use timers. For instance, if your window can be raised or lowered in eight seconds, but the timer is set for 15 seconds, that's seven seconds the window motor has power, but the window is already up or down. That's like trying to force the knob down on an overheated toaster—sooner or later the window motor will be fried.

Getting to the window motors and door lock actuators, which reside inside the door shell of your Mustang, requires the removal of your door trim panels. These door trim panels are retained by a few well placed screws and a bevy of plastic door panel retainers. After removing the armrest and the select few screws, use a door panel tool to detach the plastic door panel retainers from their stamped holes in the door shell. The use of this tool will prevent door trim panel damage and the loss or damaging of the door trim panel retainers.

Once the door trim panel is removed, a thorough inspection of the window track and regulator needs to be made to ensure the problem you are experiencing is indeed a bad window motor and not a failed track roller or a broken spot weld on the window regulator. This may require the removal of the motor and manual operation of the window up and down by a helper while you inspect the inner workings. This statement holds true for the door lock actuators as well. If the lock won't move, you may have a bent lock rod or damaged latch mechanism. Remove the lock actuator and try the lock mechanism again.

Replacement power window motors and power door lock actuators are available from your Ford dealer, as well as many aftermarket sources, like Mustang specific parts outlets. Ford, as well as some large parts store chains, carry replacement window motor gear kits that keep costs down if the motor is good, but the gear is stripped.

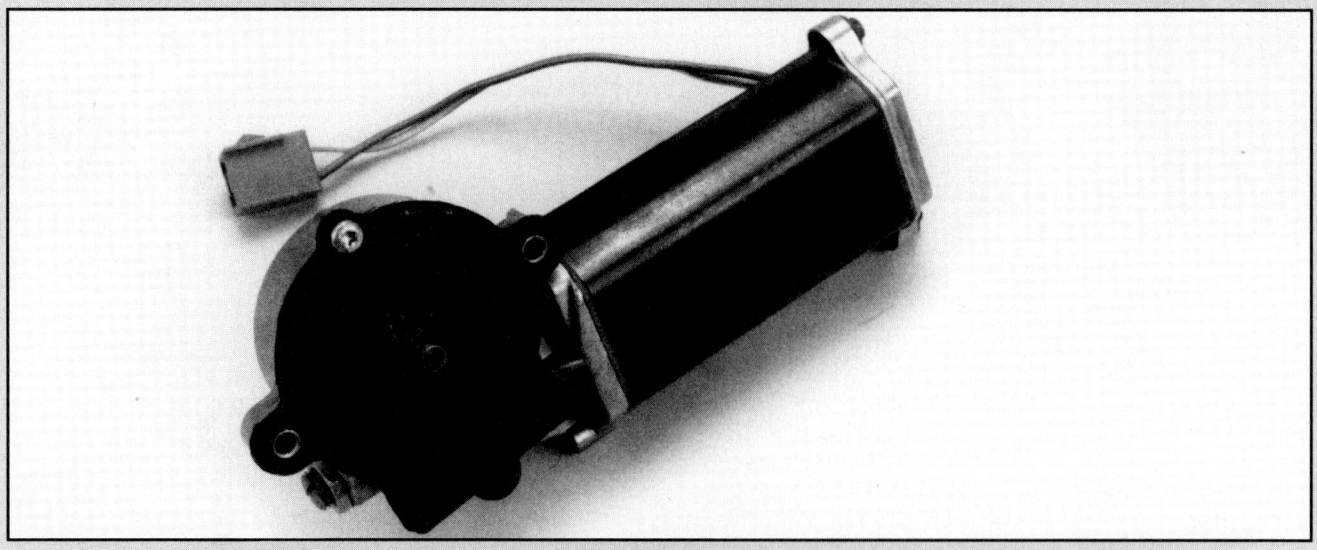

The Mustang door window motor uses a short nine-tooth steel gear to drive the window regulator. There is a left and right specific motor, so be sure of what you are ordering.

POWER WINDOWS & DOOR LOCKS

1. On convertible Mustangs with power quarter windows, the motor uses a nine-tooth steel gear, also, but it is extended on a shaft to reach the regulator within the top mechanism. The rear motors are also side-specific.

2. Our '88 GT convertible had two inoperative door locks, both rear quarter windows slipping, and the driver's side front window inoperative. We started with the master switch on the driver's door. Using a wiring diagram, such as the Ford Electrical and Vacuum Troubleshooting Manual (EVTM), check the switches one at a time in both up/down and lock/unlock modes with either a volt/ohm meter or a test light. All our switches checked OK.

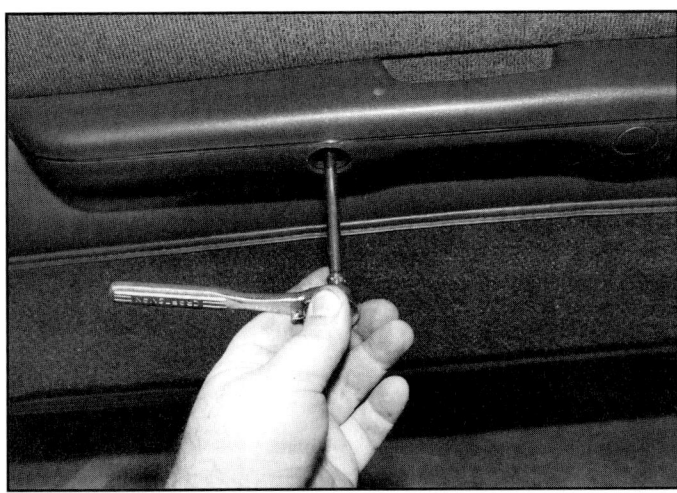

3. Since the switches were fine, we would have to go into the door for further diagnosis. Begin by removing the armrest assembly from the door panel. There are two 11 mm bolts behind the cap plugs, and two Phillips head screws, one at each end of the arm rest. Carefully slip the control panel through the armrest and set the armrest aside.

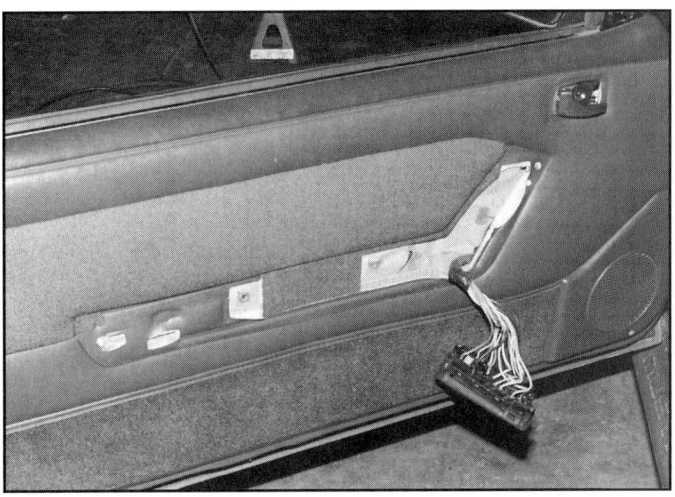

4. Once the armrest is removed, you will have to remove the outside mirror trim bezel, door handle bezel, and the speaker grille retaining screw in the lower right corner before pulling the panel free with a panel tool.

POWER WINDOWS & DOOR LOCKS

5. Using the same test light or volt/ohm meter as before, check for power at the window motor connector, again in both the up and down side of the connector. Ours checked out fine, and turned out to be the window motor itself. The motor was replaced with the removal of the speaker for access, and the three bolts which retain the motor (arrow). With the key on, use the power window switch to help you mesh the window gear with the regulator. If there had been no power, we would have known there to be a problem between the switch and the motor, and further investigation would be made into the harness to check for a broken or shorted wire.

6. Our rear windows were working electrically, but were skipping, or chattering, which led us to believe we had a bad gear in the motor, although we checked our switches to make sure. To access the rear window motors on a convertible, pull the lower seat cushion out and remove the four retaining screws which hold the trim panel in place, and pull the panel forward for access. You can see the motor behind the dust shield (arrow). Three 10 mm bolts hold the motor and motor bracket to the quarter panel.

7. With the motor out, remove the three 8 mm bolts that hold the adapter bracket to the motor.

8. Remove the one Phillips head screw and lift the gear housing cover off the motor for inspection.

POWER WINDOWS & DOOR LOCKS

9. Just as we had suspected, the nine-tooth steel gear is spinning inside the 44-tooth plastic gear. Why does this happen? The metal gear is seated inside the plastic gear with three plastic dowels that prevent the metal gear from spinning. Over time these three dowels break down and allow the metal gear to spin within the plastic gear, causing the window to go up and down erratically, or sometimes not stay up at all. If you can push your window down by hand, the gear is probably bad. Unfortunately, Ford does not sell rear quarter window gear kits, but we have found a way to circumvent that by using a front window gear kit.

10. Clean the motor of all old grease with brake cleaner or carburetor cleaner. There must be no bits of old dowel plastic in the worm drive area or the motor might bind. Shown in front of our cleaned and ready motor is a front window motor gear kit. Notice the short nine-tooth gear installed. The three dowels are under the lip of the steel gear, and are a slightly darker plastic than the white 44-tooth gear itself.

11. Here is where we discovered we could save some money and make a part that doesn't exist. Carefully pry the short gear out of the large plastic gear and set it aside. After cleaning the long shaft gear, place the gear and the three dowels into place as shown here. Next, carefully press the metal gear into the large plastic gear, while preventing the three dowels from becoming airborne. This step takes a little practice and patience, but it can be done. Once you have pressed the gear all the way in you have now built your own rear quarter window gear repair kit.

12. Slide the gear back into the motor with a light layer of grease under it, then squeeze the remaining grease into the gear housing and spread liberally with your fingers. Reinstall the gear cover and the motor adapter plate, and refit the motor to the regulator using the window switch to "bump" the gear into place.

POWER WINDOWS & DOOR LOCKS

13. Power door lock actuators are easy to diagnose and almost as easy to install. The new actuators come with new retaining brackets, but it is easier to pry the motor out of the bracket than replace the complete assembly.

14. With the door panel still off, we had only to unplug the actuator and test for power at the connector while pressing the door lock switch. Again, we had power in both lock and unlock modes for both doors, verifying we had two bad door lock actuators.

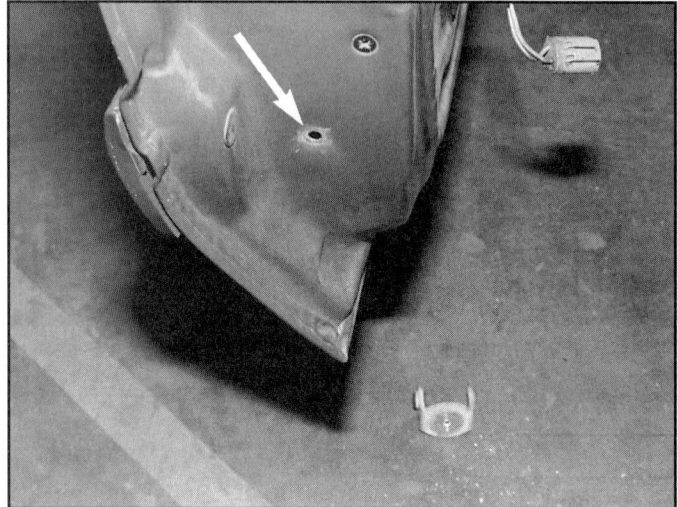

15. Sometimes all does not go as planned. While attempting to pry the actuator from the retainer, we broke the retainer, requiring us to drill out the special rivet in the door skin (arrow). The broken retainer can be seen on the floor in the foreground.

16. Since most people will not have a "Big Daddy" style rivet gun, a 1/4-inch 20 bolt, nut, and lock washer were used to retain the bracket and actuator. The passenger side didn't give us any problems, and we were able to pry the old actuator out and press the new one in by hand. We now had fully functional power windows and door locks again.

POWER TIMING

Power Timing, a concept where your ignition's timing is optimized to your particular combination, is done by starting with the stock 10 degrees of initial timing, raising it two degrees at a time and then road testing it at full throttle with each adjustment. Then, when it pings, back it off one degree. This leaves the car on the verge of pinging while giving you the best initial timing for low end response. The basic task of checking and adjusting ignition timing certainly won't take a weekend, but many of the performance projects you will come across in this book can be enhanced even further by power timing your 5.0 after the modification.

The timing of a 5.0 Mustang is critical for performance and emissions concerns. Don't set your timing at whatever your friend or track guru of the month has his 5.0 set at, since every car works differently. Adjusting your timing takes no more than an accurate timing light, distributor wrench, and a crayon or grease pencil (touch up paint will work too). Some tips and tricks to setting your timing:

• Don't forget to install the Spout connector. Those last minute timing checks at the track can get you in trouble if you forget to reinstall the jumper. How does a loss of .4 in ET for that little jumper sound?

• The dial back to zero option some timing lights have is useful to check how much advance the vacuum or PCM is giving you at a certain rpm.

• If you have a fiberglass hood, or your hood light bulb is missing, you can place the inductive pickup on the coil wire with the car running and use the timing light as a flash light to find the hold down bolt or other item under the hood.

• Computer chips have their own timing tables. The chip's tables are not in effect when the Spout is unplugged so don't worry about them.

• A Mustang running a blower or nitrous will have to run a reduced amount of timing due to the higher combustion chamber pressures. Most nitrous kits, like The Nitrous Works EFI kit, suggest an initial timing of just three to four degrees, while most superchargers like to see 6 to 8.

• Using a high octane fuel, such as 93 octane, will allow more timing to be dialed in. My own LX uses 93 octane, with 14 degrees initial timing, and we encounter no problems in the Florida heat. Manual transmission cars can somtimes go a bit higher, upwards of 16 to 18 degrees sometimes, but we suggest those numbers for track use with racing fuels, as extended street use will cause you headaches.

• If you really don't care to carry a timing light everywhere you go, a timing computer or a computer with timing capabilities, such as the Crane Interceptor or the Smart Spark units, will allow timing changes from the driver's seat under all types of driving situations, such as part throttle, light cruise, full throttle, and so on.

• The typical Mustang timing advancement to 14 degrees is good for .2 in ET.

• On nitrous cars, or if you run a high initial timing at the track, you can make two marks on the intake and one on the distributor to mark "stock" timing and "track" timing. This saves you the time of having to break out the timing light when hitting the test and tune nights.

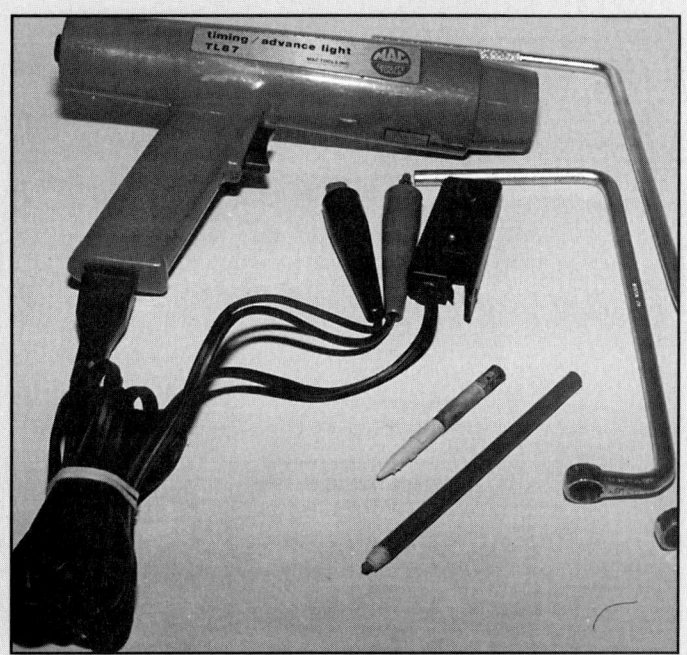

A quality timing light, grease pencil, and a distributor wrench for your type of vehicle are all that you need to check and reset your timing.

POWER TIMING

1. Determine what type of ignition system you have first. If you have a carbureted Mustang the distributor will use a vacuum advance system similar to a vintage Mustang. A Mustang equipped with EFI will have an electronic advance system controlled by the Powertrain Control Module and no vacuum advance canister will be present.

2. Thoroughly clean the crankshaft damper area that has the timing marks on it until the marks are clearly visible (accomplished with the numbers facing down and done from under the car) and then mark the crankshaft with a grease pencil or "white out" at the 0, 10, 12, 14, and 16 degree marks BTDC. Personally, I like to mark the 0 with a half length mark, the 10 with a full length mark, and the remaining two degree increments with another color. This makes it easier to determine just what amount of timing you have dialed-in to your Mustang.

3. Now that the timing marks have been highlighted you can hook up your timing light. Place the two alligator clips on the positive and negative battery terminals, as per the light's instructions (left), and then place the inductive clamp around the number one plug wire (right).

POWER TIMING

4. Since you have previously determined what type of ignition you have, you will either have to disconnect and plug the vacuum line (a golf tee works great for this) at the vacuum advance canister or unplug the "Spout" connector. Spout stands for SPark OUTput in the EEC-IV terminology. There are two types of spout connectors. The shorting bar type jumper shown here with two wires in the plug and the male/female connector type with one wire on each side of the plug. The shorting bar jumper has been used since about 1987 but the male/female type was used on other Ford products and some four cylinder Mustangs, including turbo models. The Spout connector will be located about six inches down the harness from the TFI ignition module connector on the distributor. Disconnect the Spout signal or vacuum line and start the car, being careful of the timing light's cables.

5. Set the "dial-back-to-zero" knob, if so equipped, to zero and check the base timing of your engine. Believe it or not many Mustangs come right from the factory with less than the specified 10 degrees initial timing. I have seen initial timing set as low as six degrees from Dearborn.

6. Using the distributor wrench, loosen the distributor hold-down bolt just enough to allow slight movement of the distributor. Turn the distributor clockwise to advance and counter-clockwise to retard the initial timing. When you have reset your timing, around 12-14 degrees, tighten the hold-down bolt back down and reinstall the Spout connector or vacuum line and take the car for a full throttle test drive. If the car pings try backing the timing off one or two degrees at a time until the pinging goes away.

HVAC REPAIR

After several years of use and unintentional neglect, the cooling and air conditioning systems on a late model Mustang scream for attention. Sometimes these problems go unnoticed. Visual problems, such as swollen hoses, coolant under the passenger side of the dash, the windshield fogging up when the car gets hot, and others, are all cries for help from your cooling system. The air conditioning system is less subtle. The system has a leak somewhere, the R-12 leaks out, and then you no longer have cold air conditioning. That in itself should wake you up to a problem.

This particular example of a lack of maintenance we have here is an '87 Saleen Mustang. The air conditioning hasn't worked in over a year, the thermostat sticks, the heater core leaks, and most of the original eight-year-old hoses are still under the hood. What makes these problems such a nightmare is the location of some of the failed parts. The air conditioning evaporator and heater core are inside the HVAC case behind the dashboard, a frightful thought for some do-it-yourselfers. The labor alone for such a repair usually runs several hundred dollars, not to mention the parts. What most people don't realize is that the only special tools needed are a set of spring lock coupler tools (like the ones used for fuel lines) that can be had for around ten dollars. We figured we could save some money on labor by doing it ourselves, and have the satisfaction of knowing we did it. The only time you will need a professional shop's assistance is with reclaiming and filling of refrigerant for your air conditioning system.

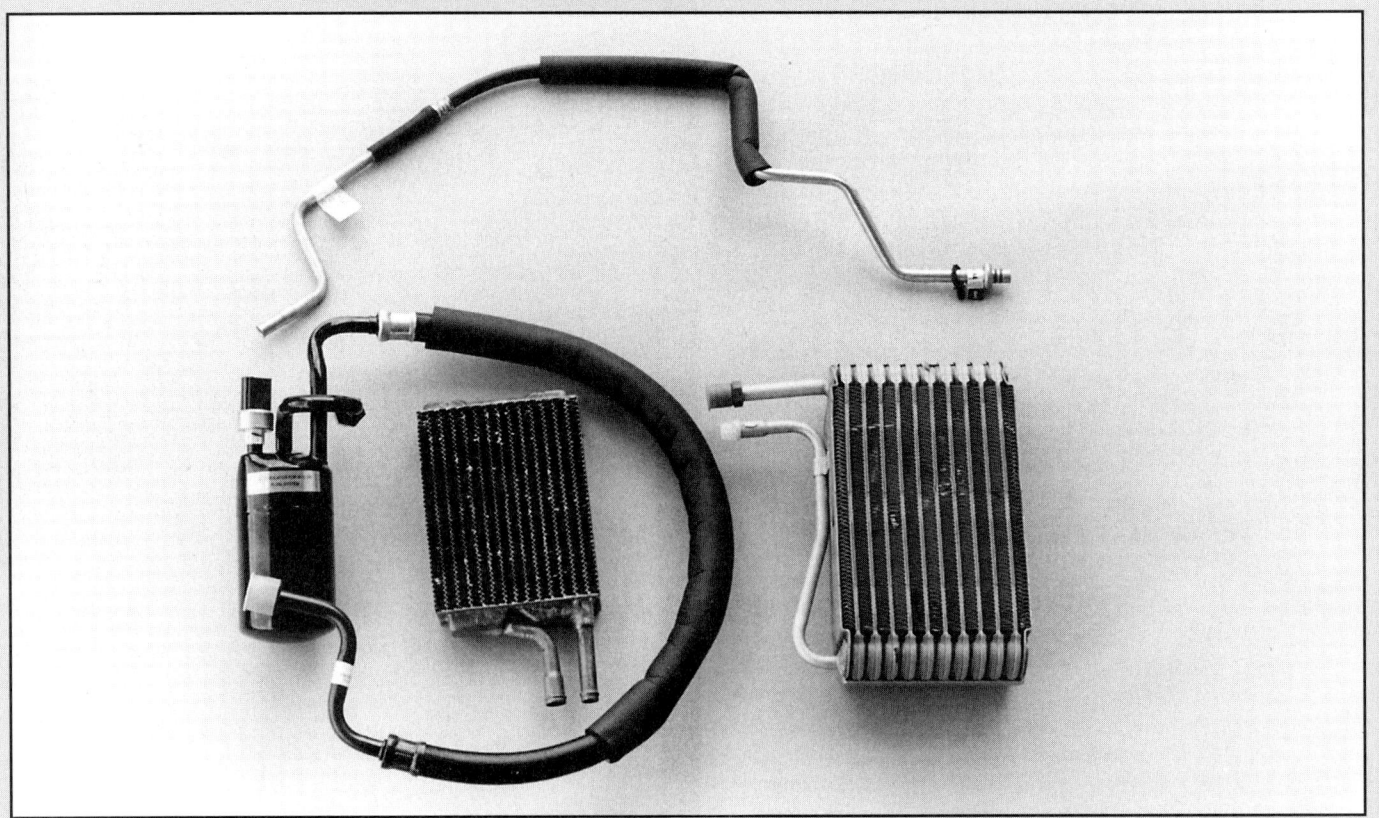

The evaporator, dryer assembly (with attached hose), the fixed orifice tube (or liquid line), heater core, new hoses, thermostat, coolant, and clamps made up our shopping list.

HVAC REPAIR

1. This '87 had no refrigerant left in it, but if your system still has some refrigerant then you will need to have it professionally reclaimed. Disconnect the battery and begin your work at the firewall. The dryer is disconnected from the compressor and the evaporator, which are simple threaded fittings, then the 8mm retaining bolt is loosened on the dryer bracket and the dryer pulled free of its home. Don't forget to disconnect the pressure switch.

2. The liquid line is disconnected from the evaporator with a spring lock tool. The spring lock tool is also used at the opposite end of the liquid line at the condenser.

3. A deep socket is required to remove the 10mm bolts that retain the dryer bracket to the HVAC case studs protruding through the firewall. The same socket is used to remvoe the large nut/washer assembly under the dryer bracket. The vacuum line coming from the firewall seal is disconnected from the one-way check valve at this time. Don't forget to disconnect the two heater hoses going to the heater core. Ours aren't shown here, as the core was leaking and the hose ends had been plugged for some time.

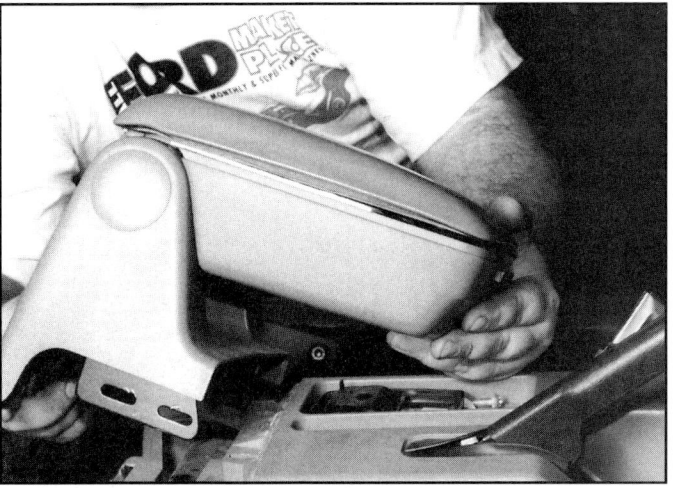

4. While not necessary to complete the job, we removed both front seats for better access and better photography. In order to remove the dash assembly the complete console will have to removed from the car. Disconnect the battery and then start with the arm rest, by removing the two side access plugs and unscrewing the four 8mm bolts and lift the arm rest free of the console.

HVAC REPAIR

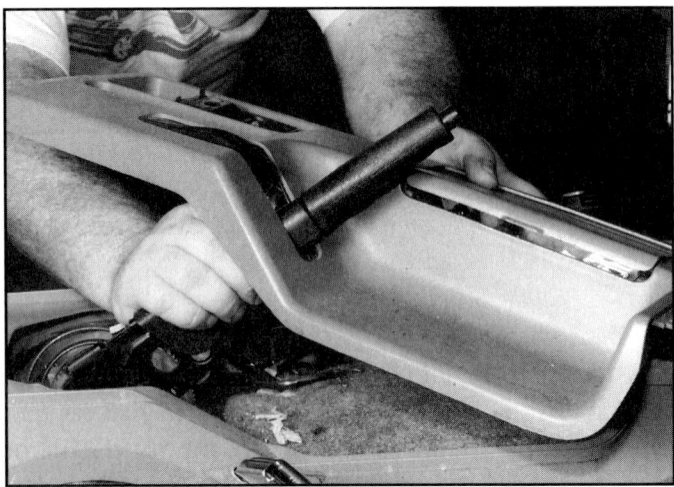

5. Four Phillips-head screws hold down the top half of the console. Once the screws are free, disconnect the electrical connections for the lighter, and the power mirrors (if equipped). Now would be a good time to fix your stubborn ash tray door spring if it isn't working.

6. There are two Phillips-head screws on either side of the forward end of the console. The driver's side is shown here. You will have to remove the lower dash panel to access these two screws.

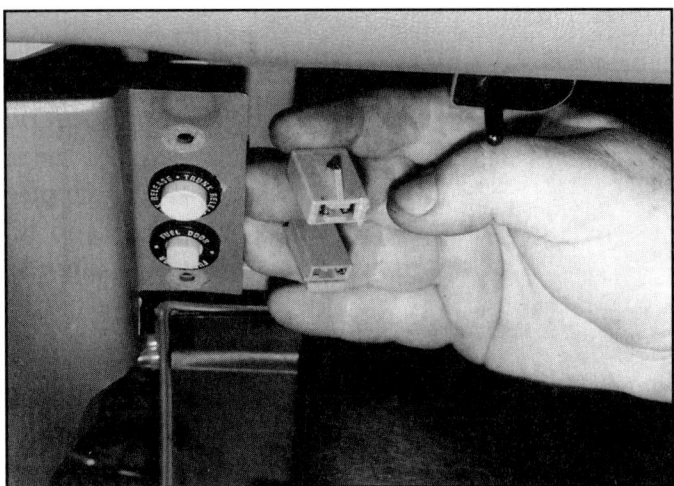

7. The passenger side screws are located behind the glovebox door. Open the door, push the sides in to release the tabs and let the door hang down. Remove the two screws from this side and unplug the fuel door and hatch/trunk release buttons, if equipped.

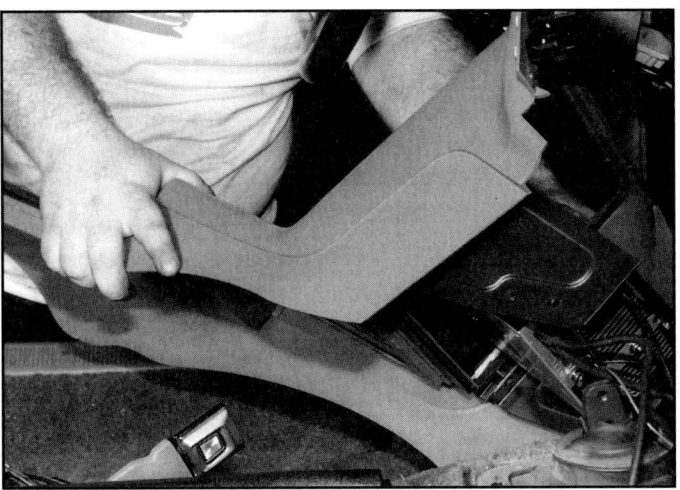

8. There are four remaining screws, two at the back of the console base, and one on each side of the stereo at the front. Once the four screws are out you can pull the console back and disconnect your stereo connections. The A/C controls stay with the dash. On five speed cars you may have to remove the shifter handle, depending upon the type of shifter you are using.

HVAC REPAIR

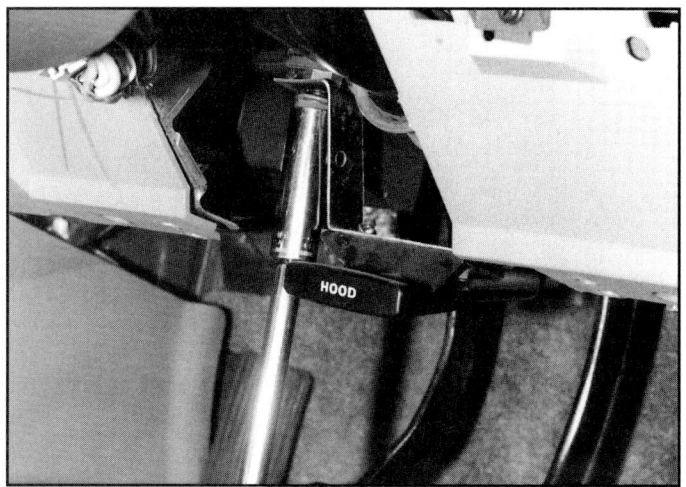

9. The steering column will need to be lowered in order to swing the dash away from the firewall. Two 15mm nuts hold the hood release bracket to the steering column brace. Once this bracket is lowered out of the way, remove the four 15mm nuts that hold the steering column to the column brace and carefully lower the column down. Supporting the column at the steering wheel with a block of wood would be a good idea. You may need to partially remove some of the column trim panels and/or disconnect the turn signal and ignition switches to prevent connector or wiring damage.

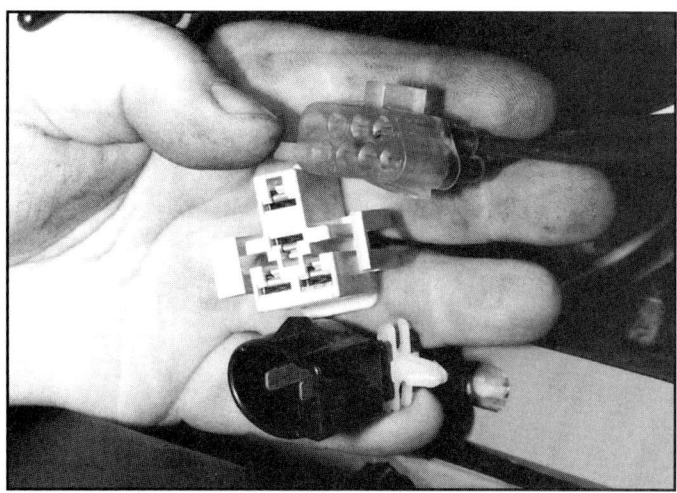

10. Return to the passenger side of the dash and reach through the glove box opening and disconnect these three items: the blower motor harness, the resistor plug at the HVAC case, and the vacuum tube harness.

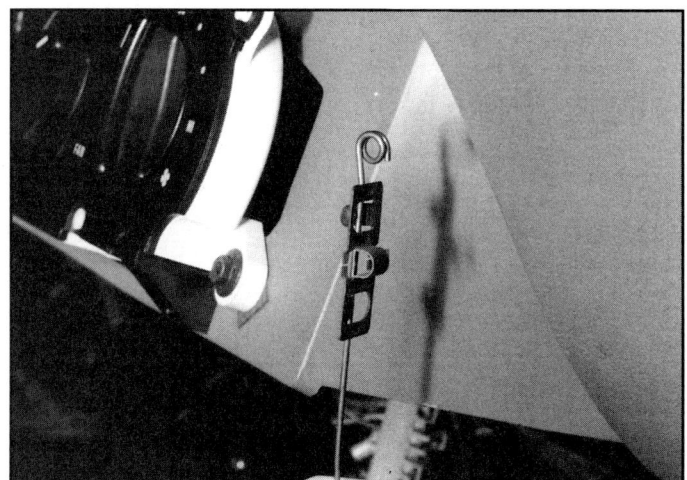

11. The temperature blend door cable is shown here. Simply squeeze the plastic tabs to remove it from the side of the HVAC case and slide the metal clip off the actuating rod. Though we unhooked this cable now, you can wait until you pull the dash away as the cable is long enough to allow you to do so.

12. There are two 8mm bolts that are located at the lower corners of the dash. You will need to remove the kick panels to access them. There is one 10mm nut directly under the gauge cluster area where the steering column resides; don't miss that one or the dash won't budge.

HVAC REPAIR

13. Finally, pry out the defroster vent trim at the windshield and remove the two dash speaker grilles to access the five 7mm bolts along the top of the dash. At this time the dash should practically slide down and into your lap, but a gentle lift and pull away from the firewall may be needed.

14. Three 10mm bolts hold the HVAC case to the firewall, two at the top shown here, and one self tapping bolt at the bottom of the case where the case and the carpet edge meet.

15. Once all the bolts are out, carefully pull the HVAC case away from the firewall and out of the car. A friend might be at the ready to hand the case off to him.

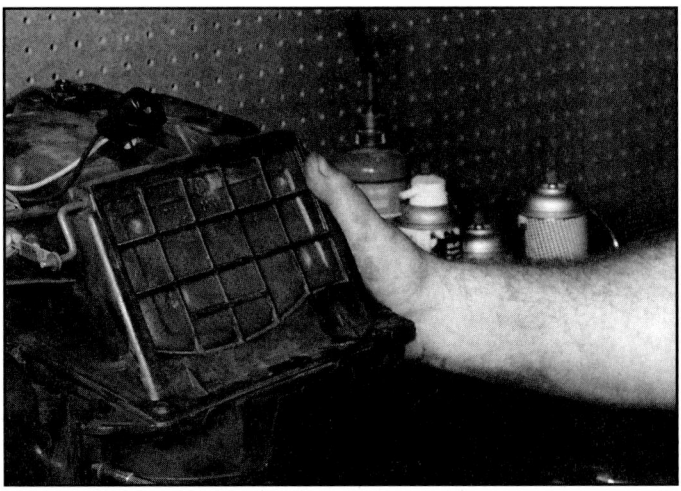

16. Continuing our work on a workbench, the blower motor case section is removed to reduce the size of the case for ease of work. Two 8mm bolts hold the cases together. Once these two bolts are removed the blower case can be tilted away from the rest of the case.

HVAC REPAIR

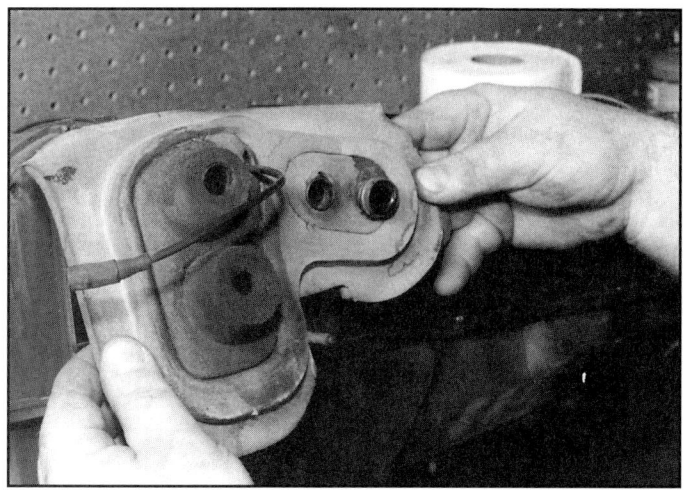

17. Remove the large firewall seal from the heater core and evaporator nipples. Be careful not to damage the vacuum line that runs through the seal.

18. Four 8mm bolts hold the cover down on the heater core. Remove the bolts and the cover to access the heater core.

19. It is plainly visible that we had several pinhole leaks in our heater core.

20. The evaporator is installed as the case is assembled from the factory, therefore, there is no lid to remove for access. You have to cut the case open, though Ford does give you an outline of where to cut. A hacksaw or other cutting tool will suffice, but we used a cut-off wheel for speed.

HVAC REPAIR

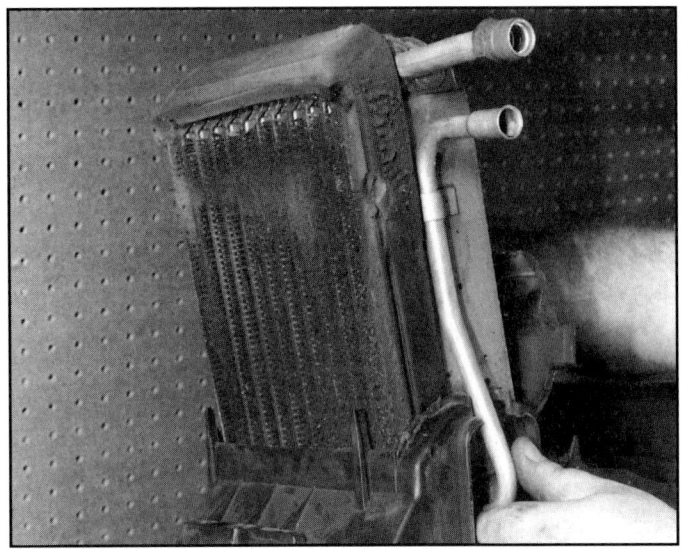

21. Bend the "door" back out of the way and pull the evaporator free of the HVAC case. Notice all the grunge covering the evaporator coils, reducing air flow and efficiency. The refrigerant oil leaking out attracts the dirt and it collects on the coils. Now is a good time to clean out the case of any leaves, coolant, trash, and other items, and check the A/C drain to ensure it is unplugged.

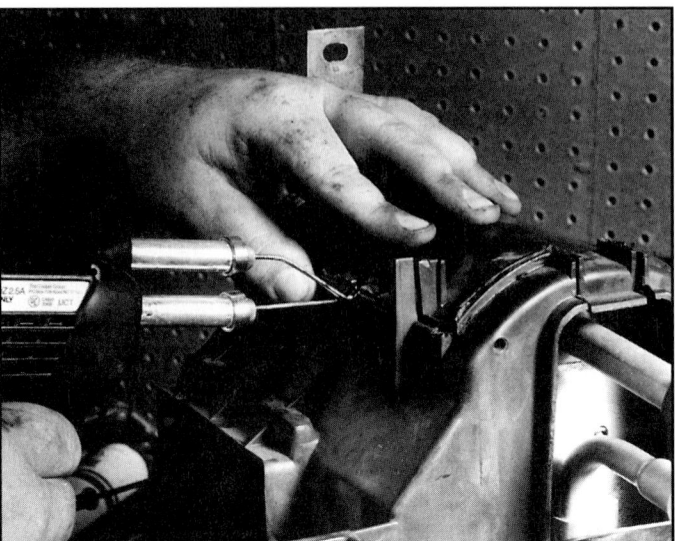

22. New evaporator seals are available from your Ford dealer, but we reused ours with some trim adhesive. Carefully insert the new evaporator into the case and bend the door back shut. Ford says to install J-clips over the two tabs and run a screw through them to keep the door closed, but the best way we could think of to keep the door closed was to grab a soldering iron and melt the plastic back together. This operation worked fine for most of the cuts, but some were too large, necessitating some strip caulking over these areas.

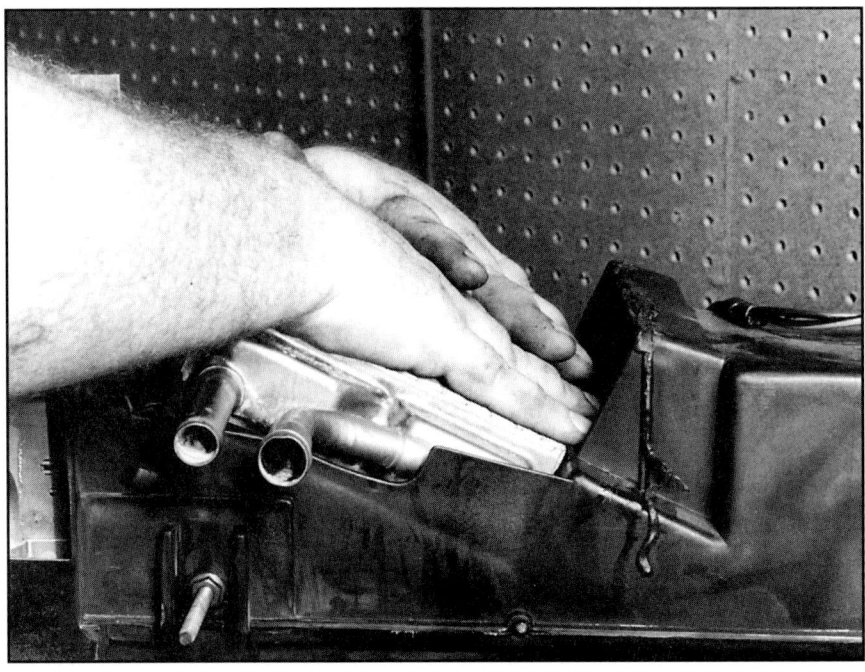

23. Lay new strip caulking in the heater core area and press the heater core into the caulking.

HVAC REPAIR

24. Add caulking around the heater core lid as needed and bolt the lid back into place. Reinstall the blower motor case half to the HVAC case half.

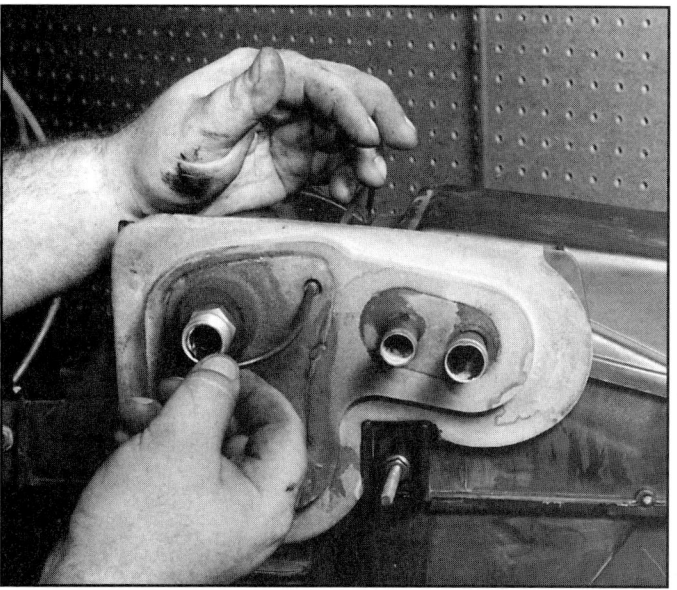

25. After installing the firewall seal, place the vacuum line inside the evaporator tube as shown. This will prevent damage to the line as you install the HVAC case back behind the dash. Reverse procedures 1-15 to complete the evaporator and heater core replacement.

26. Installation of the new dryer and liquid lines is the same as removal, just make sure you lubricate the O-rings with some assembly oil to aid in installation and sealing.

HVAC REPAIR

27. Our condenser line O-rings, and the remaining O-ring on the compressor discharge line, were all replaced from an auto parts store kit such as this. You can simply remove your O-rings and match them up at any well equipped parts store.

28. Most parts stores will carry the upper and lower radiator hoses and the thermostat bypass hose, but the two heater hoses, and the one next to the bypass hose on a 5.0 are usually a dealer only molded hose. They can be ordered under part numbers E7SZ-18472-A, E6LY-18472-C, and E6SZ-18472-A. The 3/4" heater hose even comes with a new coolant restricter. Owners of 2.3 Mustangs can use bulk straight hoses for your heater hoses instead of paying top dollar for the Ford ones.

29. After replacing the thermostat a can of radiator flush was used to remove as much deposits as possible. The cooling system was refilled and another trip was made to have our A/C system evacuated and recharged. A 48° center vent discharge temperature should stave off the summer heat wave with no problems. Bring on the heat, Mr. Sunshine.

REPLACING AXLE BEARINGS

You're driving to work one morning and you hear a strange noise from the rear of your car. You turn down the stereo and crook your neck a bit to get a better ear shot at the sound. It sounds like a dull thumping sound, possibly with a slight metallic ring to it. You my friend are hearing your axle bearing and rear axle crying out for help.

Early Mustangs that used the eight or nine inch axle housing assembly used a bearing pressed on to the axle. This bearing had an inner and outer race and the axle was retained in the housing, using four bolts with an attachment flange welded to the end of the axle tube. Remove these four bolts and the axle will slide out, no muss, no fuss.

The late model Mustang uses a bearing that is pressed into the axle tube end and not on the axle itself, whether it utilizes a 7.5 or 8.8 axle housing, open differential or Traction-Lok. This bearing has only an outer race, the inner race is a machined area on the axle shaft itself; not one of Ford's better ideas. The axle is also retained differently in these later cars. The axle has a groove machined into the splined end of the axle that locates a retaining clip, dubbed a "C-clip" due to the C shape of the clip. This type of axle retention makes for extra service work, as the axle housing cover must be removed and the gear oil drained to access the C-clips and allow axle removal.

Diagnosis requires raising the rear of your Mustang to place it on jack stands. Place the car in gear and allow the rear wheels to spin on the engine's power at idle. Carefully slide under the car with a mechanic's stethoscope and, being mindful of the rotating tire, place the stethoscope on the axle tube right behind the brake drum backing plate. A screwdriver tip placed in the same location, with the handle nestled in the opening of your ear (or a length of heater hose) will also work. Check both sides to determine which one is bad. If you aren't sure the sound you hear coming from the stethoscope is a bad bearing check a friend's car that does not have any problems and compare the sounds from his to yours.

Though the service entailed is a bit messy, the job can be accomplished with jack stands and basic hand tools. There is no need to remove the differential and play with shims or other measurements so the axle bearing repair is actually quite easy, giving you peace of mind and a great sense of accomplishment.

To correctly perform your axle bearing repair you will need a replacement axle bearing, an axle bearing oil seal, two quarts of gear oil (we used synthetic on our 1990 5.0), friction modifier if it is a T-Lok unit, and either cover gasket or silicone, though we prefer the gasket. You may also need a replacement axle.

1. With the rear of the car jacked up to a sufficient working height and safely supported by jackstands, remove the 10 axle housing cover bolts and gently pry the cover free. Make sure you have a suitable container to catch the draining gear oil. Set the cover, bolts, and axle ID tag aside for cleaning.

REPLACING AXLE BEARINGS

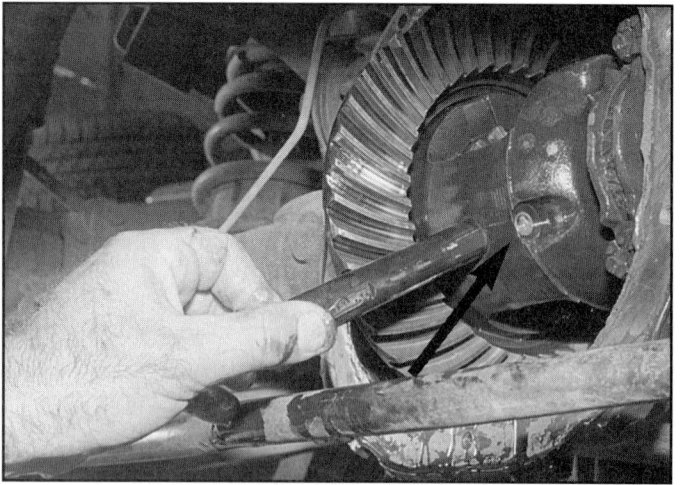

2. Clean the differential of gear oil with brake part or carburetor cleaner. Remove the differential pinion shaft lock bolt from the differential. The lock bolt is found on the passenger side of the differential, opposite the ring gear in a recess of the casting (arrow). Once the lock bolt is removed the differential pinion shaft can be dislodged from between the preload spring and the side gears.

3. Rotate the differential so that the preload spring and C-clips are visible through the service opening. Push the axle inwards, or towards the center of the car. The axle will move inwards about 3/8 of an inch or more, allowing the C-clip to be removed from the end of the axle. The use of needle nose pliers is helpful in removing the C-clips.

4. With the C-clip removed the axle can now be removed from the housing by sliding it out of the housing until the end of the axle is free.

5. Though the Ford Mustang service manual states to remove the axle bearing and oil seal together, we removed the seal first so you could get a better look at the bearing slide hammer extricating the bearing. This is the only tool we had to rent, but even a body shop slide hammer can be used if slightly modified.

REPLACING AXLE BEARINGS

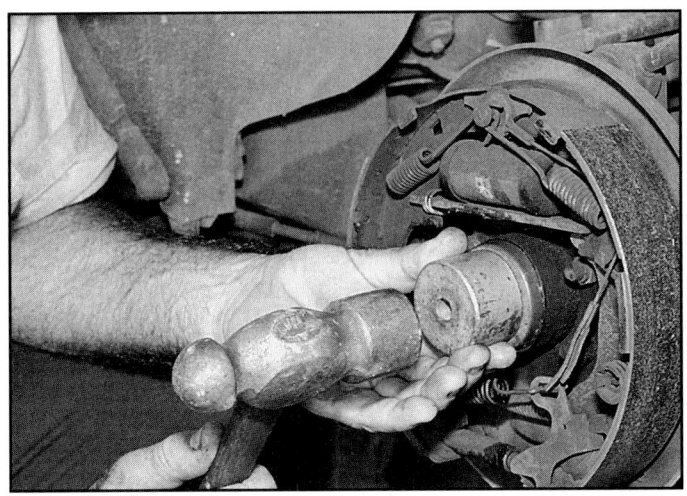

6. Install the new bearing and seal separately with either a rented installation tool, or as we did here, carefully with a large socket the same diameter of the seal and bearing. Apply gear oil or wheel bearing grease to the bearing and seal to aid in installation of the axle.

7. Since our bearing had failed some time ago, the bearing took out the axle shaft with it, which is actually quite common since by the time the human ear hears the problem it is usually too late for the axle. The marks on the machined portion of the axle are from the failed bearing which ruined the axle. While a standard 28 spline axle will set you back over a hundred dollars, we bartered a lunch for a good used one from a friend's Mustang who had upgraded to 31 spline Ford Motorsport components at barely 15,000 miles. This made for a great deal and the axle should last as long as a new one with zero miles on it.

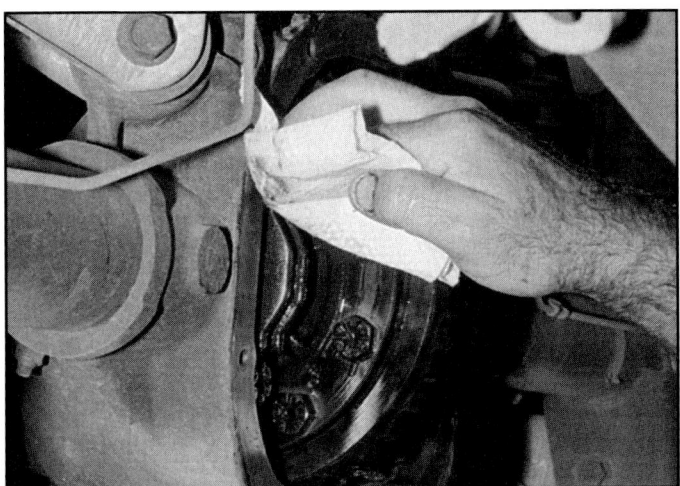

8. Once the axle is placed back into the housing, reinstall the C-clip, the pinion shaft, and the pinion shaft lock bolt, tightened to 15-30 Lb-Ft. Clean the housing cover and gasket surfaces and install the cover with the factory bolts. Add the friction modifier first (if required), and then top off the differential with gear oil until the oil just reaches the fill hole.

9. So what happens if you don't have a great friend with a spare 28 spline axle laying around. Or maybe you've upgraded to 31 spline axles a while ago and everyone else has 28 spline axles for sale. Enter the Motive Gear repair bearing. This bearing relocates the bearing inboard to use a new section of the same machined surface on the original axle, without any machine work, and with the same labor time as installing an original bearing. This will save the cost of purchasing new axles for at least another 50,000 miles. Contact Motive Gear at (312) 225-1550.

5.0 TUNE-UP

As the '79-'93 5.0 Mustangs head for their glory years on the road tune ups become a necessary evil. While there are many other makes of automobiles on the road that are worse to tune up, the 5.0 Mustang isn't like the old carbureted 289 engines of the '60s. Fuel injection, Electronic Engine Controls, emission controls, and cramped quarters make for a challenging afternoon of spark plug and filter servicing.

How much maintenance you will have to do depends upon your driving habits, mileage, and environment. If you bought the car used, whether from a dealer or a private owner, and you can't verify the maintenance history of the vehicle in question, you would be smart to replace everything. Replace all fluids, filters, and spark plugs, inspect all wiring, and check cooling system components for signs of age, wear, or abuse. If you have owned your car from day 1, you should know what the car needs and doesn't need.

To get your engine in top running order and looking like new will take some ambition, a day of your time, and some basic hand tools. Throw in several different cleaners and degreasers, and some detailing paints, and you'll be all set. It would be a good idea to use fender covers or sheet plastic taped over the cowl and fender areas to keep the over-spray of the cleaners and degreasers from damaging any paint or plastic parts. After you have cleaned the engine, rinse any excess degreaser off with a garden hose and then towel or blow dry as much as you can.

Our Mustang being tuned here is a daily driven '88 with 70,000+ miles rolled up on the odometer. After replacing the plugs, flushing the radiator, replacing all the hoses, checking and/or changing necessary fluids, replacing the PCV valve, distributor cap and rotor, and the plug wires, a detailed clean-up finished off the weekend and made the engine compartment look new again.

This Mustang has a K&N reuseable air filter, which we only had to clean. If you don't have a K&N filter, check and if necessary replace your stock air filter assembly. Remember, engines need air to breathe. You can buy standard Motorcraft maintenance items if you wish, or as in our case we used Ford Motorsport plug wires and Severe Service blue silicone hoses. These wires not only look great, but they work well too. They feature a low resistance, which in turn means minimum spark loss. The silicone hoses are the same ones that are optional on the Special Service (police) package Mustang. These hoses last the lifetime of the car and come complete with all clamps. When it comes down to counting pennies these hoses cost almost twice as much as stock rubber hoses, but you should never have to replace them again. Which translates into less maintenance later on in life, and I'm sure we all wouldn't mind that.

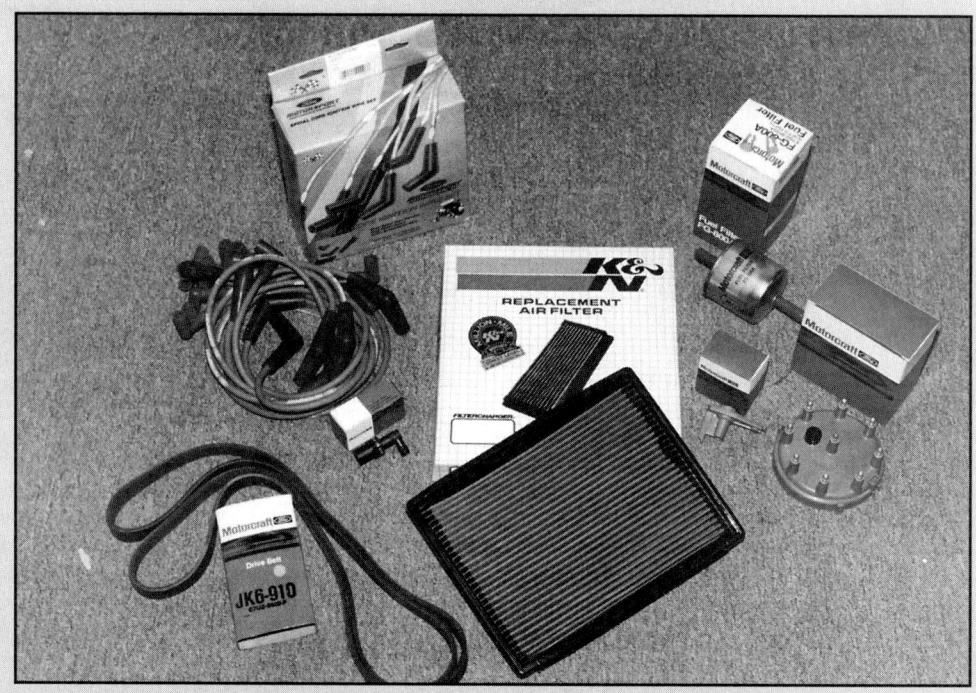

With a full complement of factory and after-market parts, we planned to clean up our '88 LX's act.

5.0 TUNE-UP

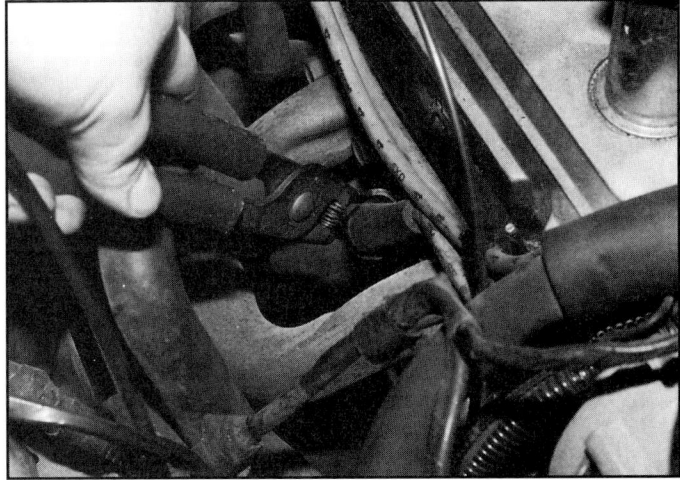

1. Start your tune up by removing the plug wires from the plugs with a pair of spark plug boot pliers. Don't worry about firing order if you are replacing your wires too, we will tackle that later.

2. Clean off the plug area with compressed air or spray solvent to prevent dirt from entering the combustion chamber and remove the plugs with a spark plug socket and ratchet.

3. The use of an old straight plug wire boot will aid in removal of the plugs from the head once they have been loosened with the ratchet. A section of 3/8" rubber hose will work too.

4. Our Split-Fire plugs hadn't seen too much service so we decided to clean, check and regap them where necessary. If you don't know what your plugs should be gapped at, check the VECI decal on the plastic coil cover at the left strut tower or check with the manufacturer of the plug.

5.0 TUNE-UP

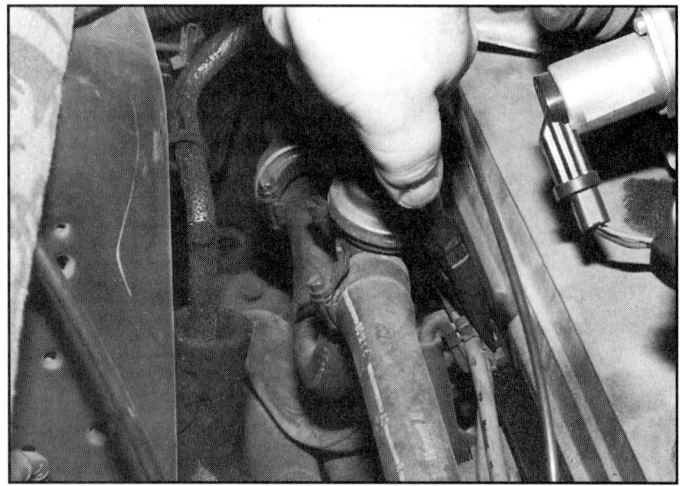

5. Using a pair of needle-nose pliers, remove these wire holders from the valve cover bolts and pull the plug wires through the various harnesses in preparation for removal.

6. After pulling all of the plug wires up and away from the valve covers, remove the plastic coil cover and remove the coil wire. Check the coil for corrosion at the tower and for arcing around the sides; if there are any signs of such, replace the coil.

7. Remove the distributor cap and all of the plug wires as an assembly.

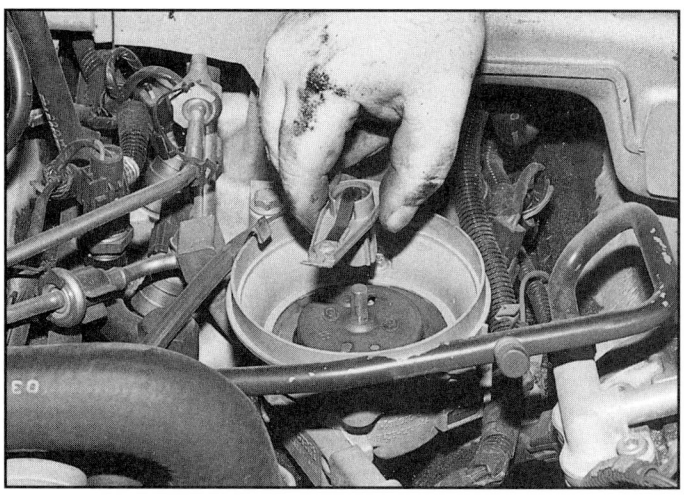

8. Don't forget to replace the rotor.

5.0 TUNE-UP

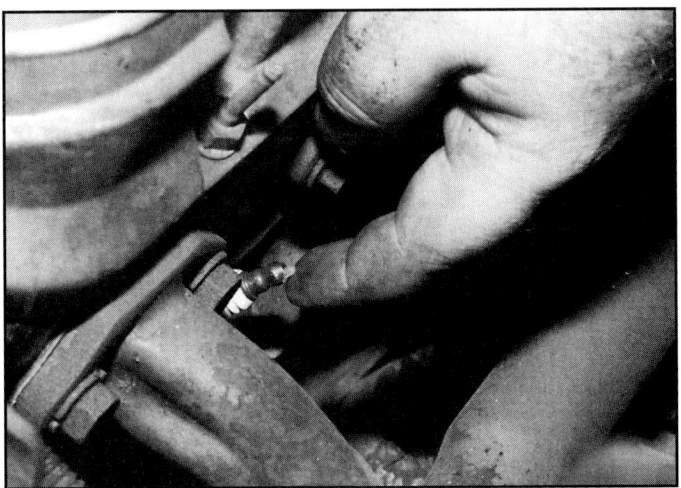

9. The new wires come with a dielectric grease that should be sparingly applied to the tips of the spark plugs, coil tower, and distributor cap towers before the new wires are installed.

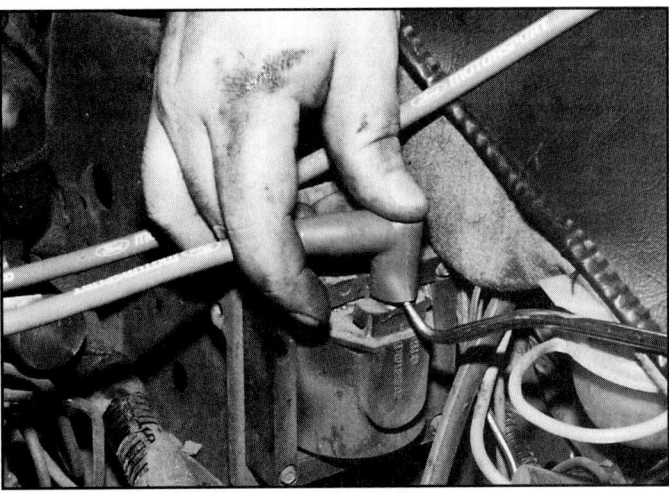

10. Install the coil wire first. Using a cotter pin remover or other blunt hook instrument, "burp" the wire boots at both ends immediately after installing the wire. Slide the tool up inside the boot and twist gently side to side until you hear a little burp of air. You will have to do this for all the wires, but only at the distributor cap end on the rest, because the spark plug end has more tension on the plug. If you don't do this the plug wires can slowly slide up and possibly fall off.

11. After you have completed the coil wire, install the rest of the wires onto the plugs (each wire is numbered for its cylinder, and the cylinders are numbered from front to back, left-1,2,3,4 and right-5,6,7,8 looking at the engine) and then push the wires onto the distributor cap following the counter clockwise firing order of 1-3-7-2-6-5-4-8.

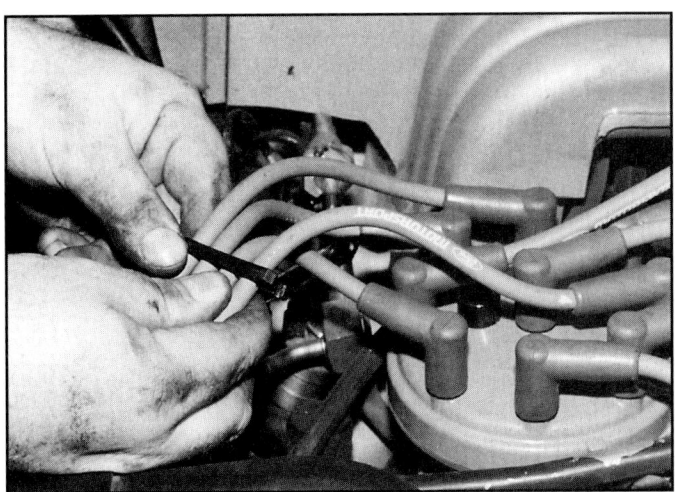

12. Place all the wires in their clips for a neat appearance. The stock wire holders can break easily so use caution.

5.0 TUNE-UP

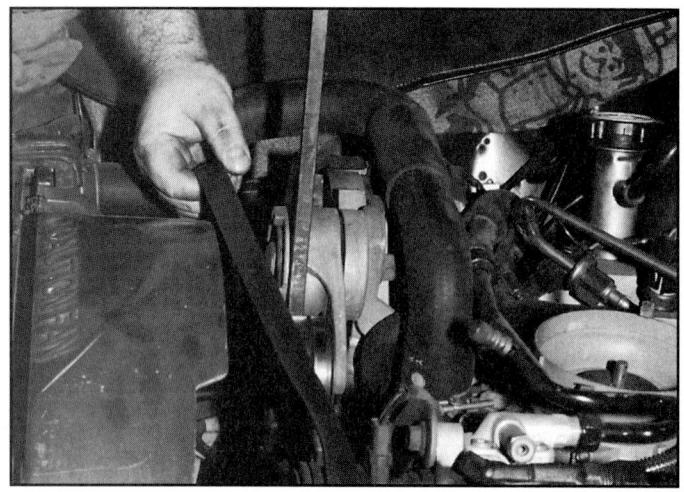

13. To replace the serpentine belt, place a long screwdriver or breaker bar between the tensioner and the pulley, and pull the handle towards you to loosen the tensioner and remove the belt. BE VERY CAREFUL when doing this, the tensioner has a tightly wound spring inside and you wouldn't want to have it slip off of your prybar and hit your hand.

14. A serpentine belt that has exceeded its life expectancy. Any severe cracks or missing chunks means it's belt time.

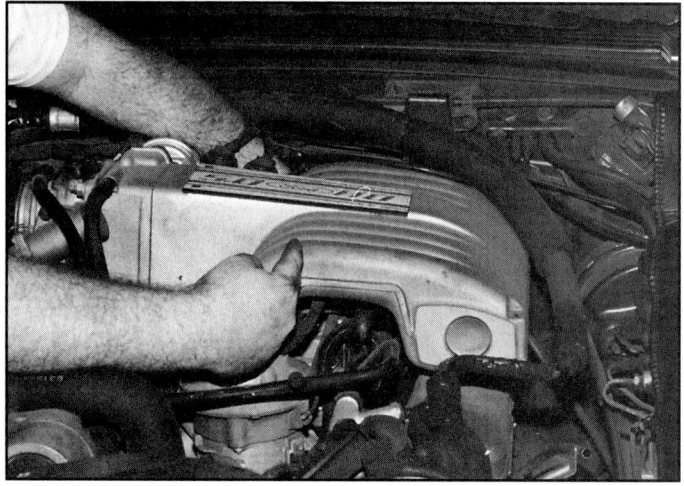

15. Just in case you forget how the belt goes back on, there is a decal showing belt location around the pulleys on the radiator support. Remember that the water pump spins backwards and that the back side of the belt, not the ribbed side, should run on the water pump pulley.

16. To replace the PCV valve, run your hand down the back of the intake in the general vacinity of the hand pointing in front and pull the PCV valve and hose out of the lower intake.

5.0 TUNE-UP

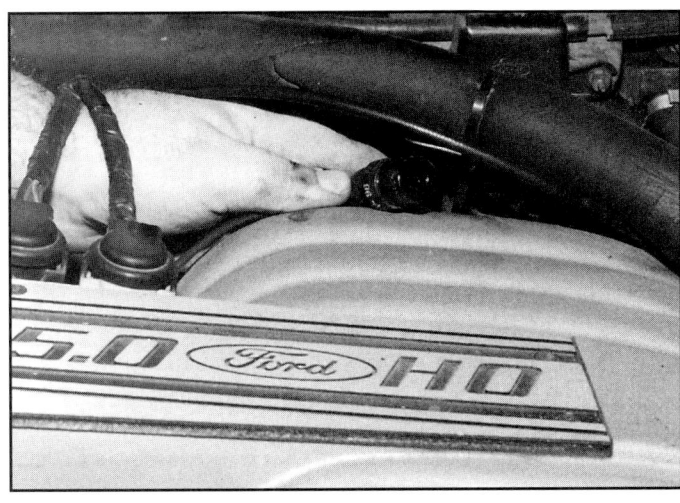

17. Here is the PCV valve after it has been snaked out of the back of the engine. Remove the valve from the hose.

18. This rubber grommet may have come out with the PCV valve; if not, reach back down there and remove it from the lower intake with the tips of your fingers.

19. Using a pick, reach back down there a third time and remove the PCV screen, part # E6ZZ-6A631-B, and check it for sludge build up. If it is hard to remove, spray some carburetor cleaner down inside it and let it soak for a few minutes. When this screen gets plugged it will cause the throttle body to build up sludge.

20. Reinstall the screen, rubber grommet, and the new PCV valve in the lower intake.

5.0 TUNE-UP

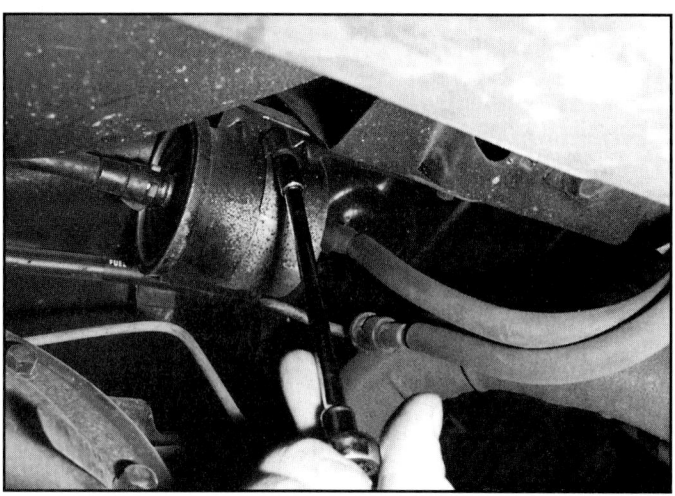

21. The fuel filter is located under the car at the rear, just forward of the gas tank and above the rear axle. Loosen the filter clamp to facilitate removal.

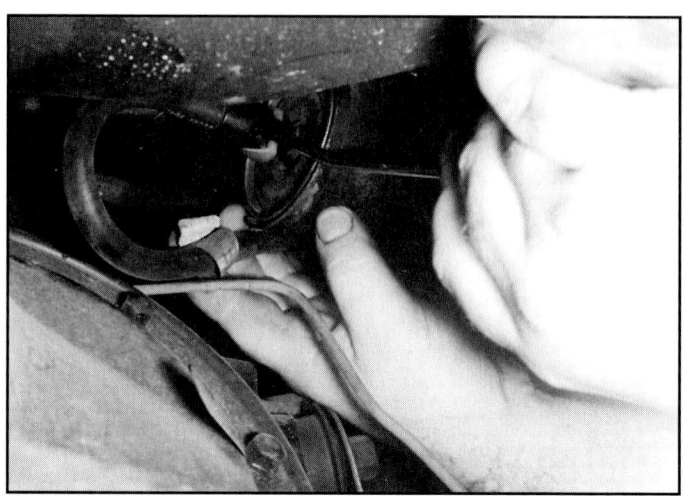

22. Remove the plastic clips from the two fuel lines and remove the lines from the filter. CAUTION: There may be some residual fuel left in the lines under pressure, so remove the lines slowly. Remove the filter, noting the direction of fuel flow, and install the new filter.

23. Hold the throttle plate open and rub your finger on the inside of the throttle bore and this is what you might find, some serious sludge buildup.

24. Spray some carburetor cleaner on a rag and wipe the throttle body out the best that you can. (Note: Do not clean your throttle body if it is a '92 or newer, or if the throttle body has a yellow sticker on it stating that the throttle body is coated.) If the throttle body is extremely sludged, removal for cleaning might be necessary.

5.0 TUNE-UP

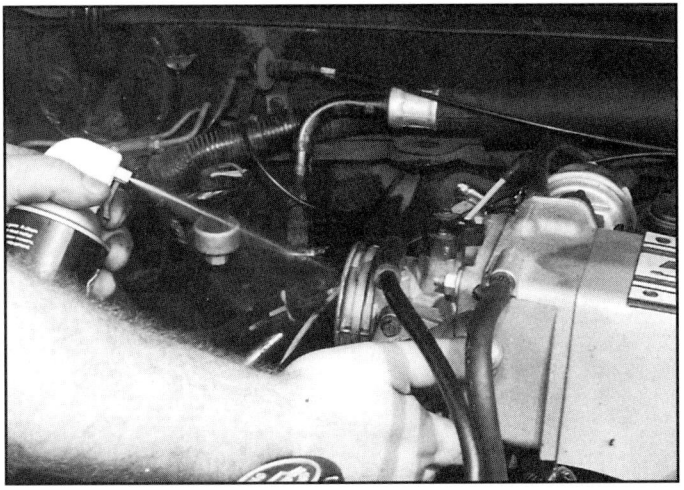

25. With the engine running, spray carburetor cleaner into the throttle body until the engine labors and then give the throttle a quick snap to clean it out. Repeat this a few times to clean the throttle body out completely. This step cannot be completed on Mass Air equipped cars so clean those models only with the rag method.

26. Remove the battery terminals and clean the battery posts and the terminals and then reassemble. Add felt anti-corrosion washers, or an anti-corrosion spray, to the terminals.

27. After a thorough cleaning, coat the terminals with a battery sealer to prevent future corrosion.

28. As you can see in this photo, our '88 had some serious grunge buildup to be taken care of. The A/C lines were coated in oily dirt and had the paint chipping off and the valve covers had oil residue on them—just a total mess.

5.0 TUNE-UP

29. Hose the engine down and get everything wet, then spray a good mix of soap and water over all the components and let it soak for a minute or two.

30. We sprayed some of the really grimy areas with engine degreaser before we rinsed the engine off.

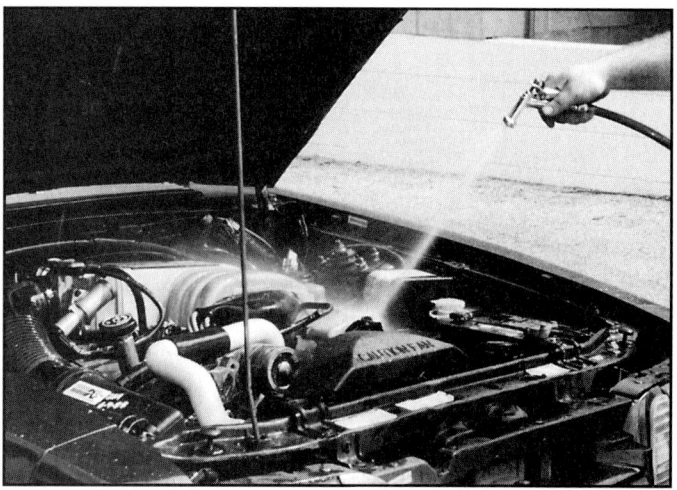

31. Rinse the engine off completely and then towel and blow-dry the engine.

32. After the engine had dried we sprayed some fresh black paint on our a/c lines and applied a coat of Armor-All to all the underhood rubber components, and low and behold there appeared a clean engine.

III. ENGINE TECH

CAT-BACK EXHAUST SYSTEM
SHORTY HEADERS
MAF METER SWAP
ROLLER ROCKER ARMS
LOW TEMP THERMOSTAT
THROTTLE BODY & EGR
K & N FUEL INJECTION PERFORMANCE KIT

CAT-BACK EXHAUST SYSTEM

When it comes to driveway projects, there is little that can compare to installing a new "cat-back" exhaust system for your 5.0 Mustang. A complete exhaust system, from the catalytic converters rearwards, can be installed easily in an afternoon with hand tools and gives your Mustang that authoritative exhaust note that the Mustang so richly deserves, but the factory denies.

While many a shade tree has completed the occasional tune up or brake job, few decide to actually work under their Mustang on their exhaust systems. But aftermarket exhaust systems and components have come a long way from that old 69 dollar pair of headers that took a trip to the muffler shop to hook up to your OE pipes 10 years ago. Today's aftermarket exhaust components are designed with the enthusiast installer in mind with pre-bent pipes, slip fit joints, and OE type flanges.

Over the years, and after installing dozens of these exhaust systems, I have found that there is actually more work involved in removing the OE exhaust components than there is in installing the aftermarket system. The original exhaust on your Mustang is a complete jig-welded system. The muffler, muffler inlet tube, and tail pipe are welded together and laid over the axle housing along the assembly line. If you wish to remove the stock exhaust without cutting (the most labor intensive removal choice) you will have to remove the rear wheels, unbolt the rear shocks, and remove the rear coil springs to have enough room to remove the exhaust as one part. On the other hand, a saws-all or hack saw will make short order of your stock parts and get them off the car without having to touch any of the rear suspension. Once your new exhaust system is in place you can either leave the clamps alone or have a muffler shop or other competent welder permanently weld the system together, just like the factory, and have the clamps removed.

Complete cat-back systems are available from Walker Dynomax, FlowMaster, Borla, MAC, Thrush, Edelbrock, and Hooker with prices ranging from just under 300 dollars to around 700 dollars, depending upon materials, pipe diameter, and shipping. One thing we don't advise is mixing components from different companies. Many of these companys' systems are designed to fit themselves and nothing else. Different bends, inside and outside diameters, material thickness, muffler and pipe length, as well as attachment points can and will differ. Save yourself from a real headache and install a complete system from one manufacturer. You should also ask owners of other 5.0 Mustangs what exhaust they have on their Mustang and "experience" the system with a short cruise. There is nothing more aggravating than spending a paycheck on an exhaust system, and then have it drive you crazy with resonance or too deep an exhaust note. The system shown here is the Walker Dynomax '87-'93 2-1/2" LX system with polished tail pipes, perfect for that stock look. The following photos show the installation on a '90 LX 5.0, but these instructions can also be used on '94-'95 Mustangs too.

CAT-BACK EXHAUST SYSTEM

1. If you wish to save the original cat-back system it will need to be removed without cutting the tailpipes from the mufflers. To do this you will need to allow for adequate clearance for removal. Place your Mustang on jackstands. Remove the rear wheels and unbolt the rear shocks, the quad shocks, the upper control arms, and remove the c-clip holding the rear brake hose in its bracket. Make sure there is a floor jack under the axle's center section. Once all four shocks and the upper control arms have been disconnected from the axle housing, lower the housing with the jack. If you wish to just cut the system apart, once on jackstands, simply unbolt the rear shocks for some extra working room and let the rear axle hang from its control arms. Cut the tailpipes where they enter the mufflers.

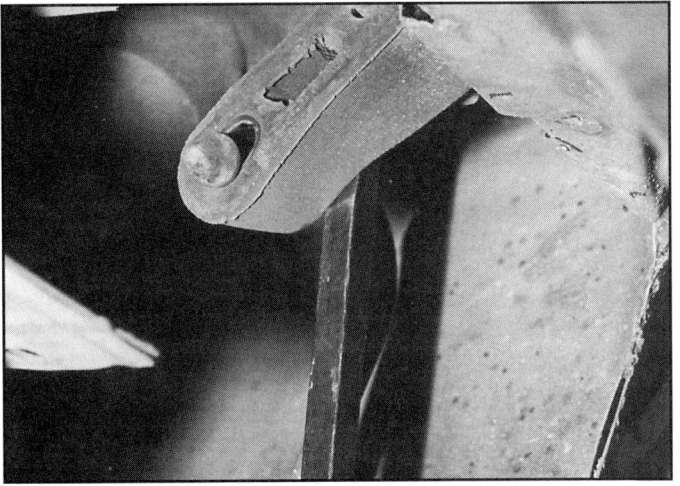

2. Pry the tailpipe out of the rubber isolator just behind the rear bumper cover. A liberal shot of silicone spray or WD-40 will help if the bushings are dry or hard.

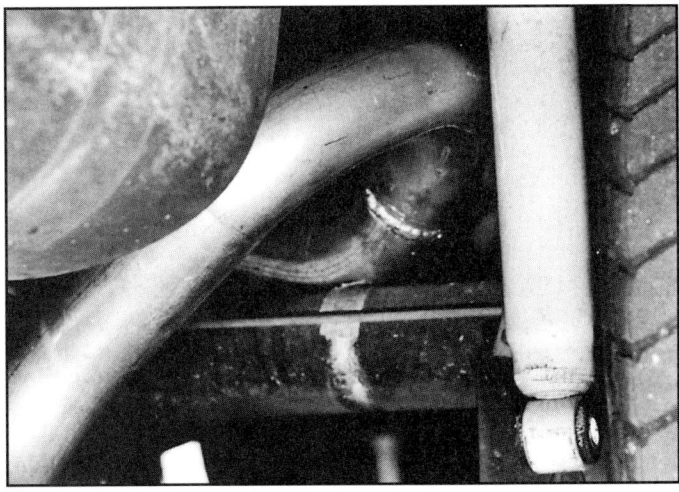

3. If you cut your system, you can remove the tailpipes from your Mustang now.

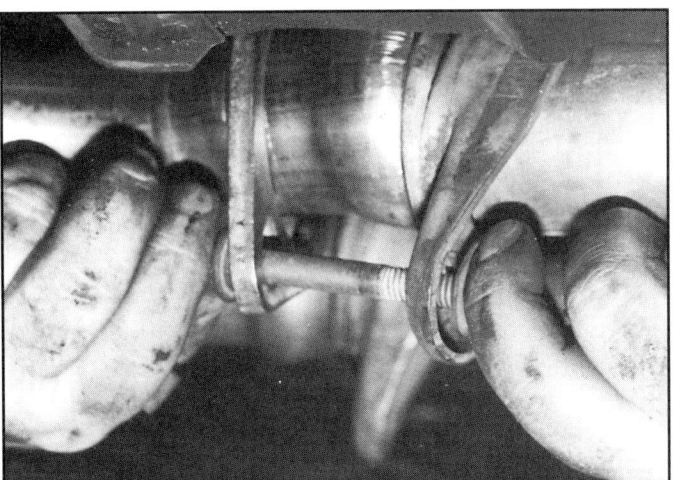

4. Working from the side of your Mustang, unbolt the intermediate pipes from the catalytic converters, then move towards the rear of your Mustang and remove the two bolts (per muffler) that retain the muffler to their hangers. Once this is accomplished you can carefully snake the muffler and tailpipe assemblies from your 5.0. If you cut your system, the muffler and intermediate pipe assemblies can be removed from under your Mustang. You are now ready to install your new cat-back system.

CAT-BACK EXHAUST SYSTEM

5. The Walker system differs from the factory system because you will have to install the intermediate pipes into the muffler and clamp them in place. This extra step might help in alignment of the system, if needed. Install the intermediate pipe into the muffler and install a clamp finger tight. Repeat this for the other muffler and pipe. Take notice of the difference in length between the two different pipes. The long one is for the passenger side and the short one is for the driver's side.

6. Place the new tailpipes over the rear axle and re-secure them to the rubber isolator just behind the bumper cover. Some silicone spray or white grease might help here again. The Walker pipes mimic the factory pipes with a small nipple on the end to prevent the pipe from dislodging.

7. After the tailpipes have been laid over the rear axle, reconnect the control arms (if disconnected) and shocks. Do not tighten the control arms until the suspension is fully compressed by a floor jack for proper ride height and quality.

CAT-BACK EXHAUST SYSTEM

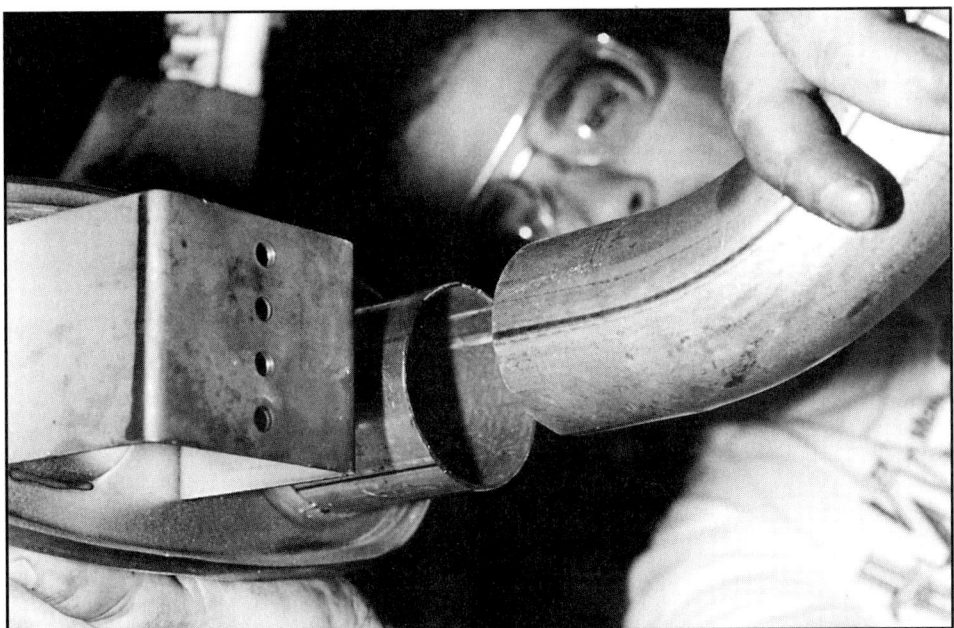

8. With the tailpipes in place you can now continue on to the mufflers and intermediate pipe installation. Simply slip the mufflers onto the tailpipes and install exhaust clamps at the muffler/tailpipe union finger tight.

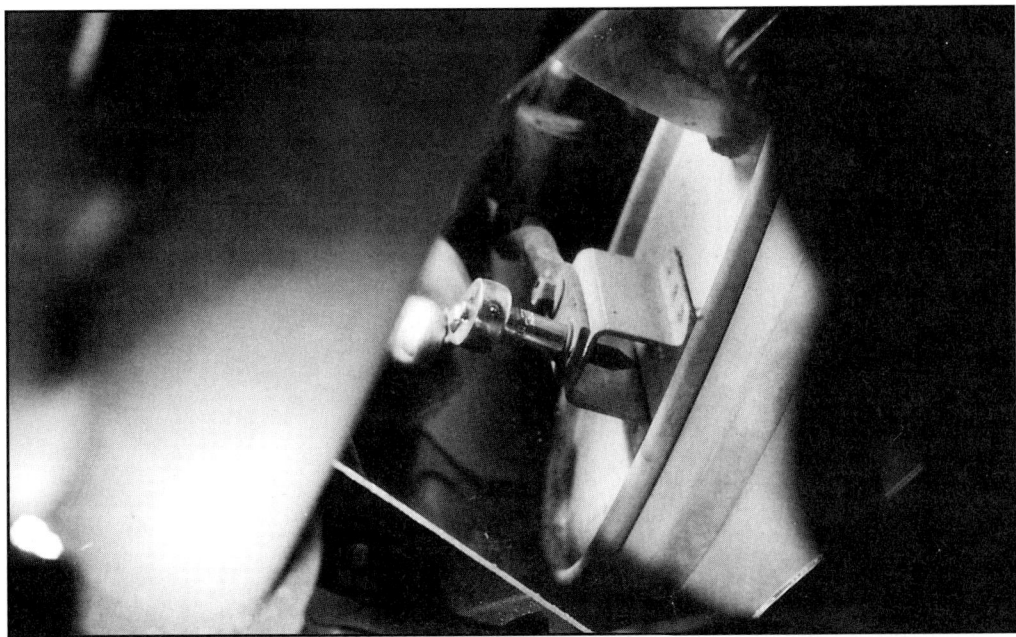

9. Bolt the isolator bracket to the rear of the muffler, using the factory hardware. Check for any possible "grounding" of the exhaust system to the underside of your Mustang that could rub through a brake or fuel line, or cause a vibration or other noise. Align the system starting at the catalytic converter to intermediate pipe flange and work your way back, including the intermediate pipe to muffler, and muffler to tailpipe. Carefully tighten the exhaust clamps as you work back.

SHORTY HEADERS

Shorty headers, a term right from the 5.0 parts dictionary. Not so long ago, there wasn't even any discussion over equal length versus unequal length, or short tube versus long tube. Instead, you would buy a set of "Hooker" or "Blackjack" headers, which had 30-plus inch long primary tubes that were designed the same way every other header since the '50s had been designed: three or four bolt collector flange with a gasket, lousy ground clearance, flat black paint that flakes off the first time you get them hot, etc. To install headers like these meant to cut the converter H-pipe or Y-pipe and a trip to the muffler shop to weld everything back together.

When the 5.0 SEFI Mustang arrived in 1986 the bolt-on parts business for these cars took off like the gold rush of the 1800s. One of the prerequisites of these new computer controlled cars is that the emissions devices now included couldn't just be chucked like so many owners had in the past. Many owners were intimidated by the EFI on their Mustangs and asked the aftermarket to give them easy to install performance parts, and this is where the shorty header was born.

The shorty header was designed as a direct replacement header that would not require the mangling of the converter system, any welding, or trips to the muffler shop. The "average Joe" with his trusty Craftsman tool set could put these headers on. While they don't achieve the performance potential of a true long tube header, a good set of shorty headers, matching high flow H-pipe, and 2 1/2" cat-back exhaust will work great on just about any street car up to the 400 horsepower range.

Shorty headers are a very popular weekend project for any '79-'95 5.0 Mustang because they are great bolt-on power enhancers.

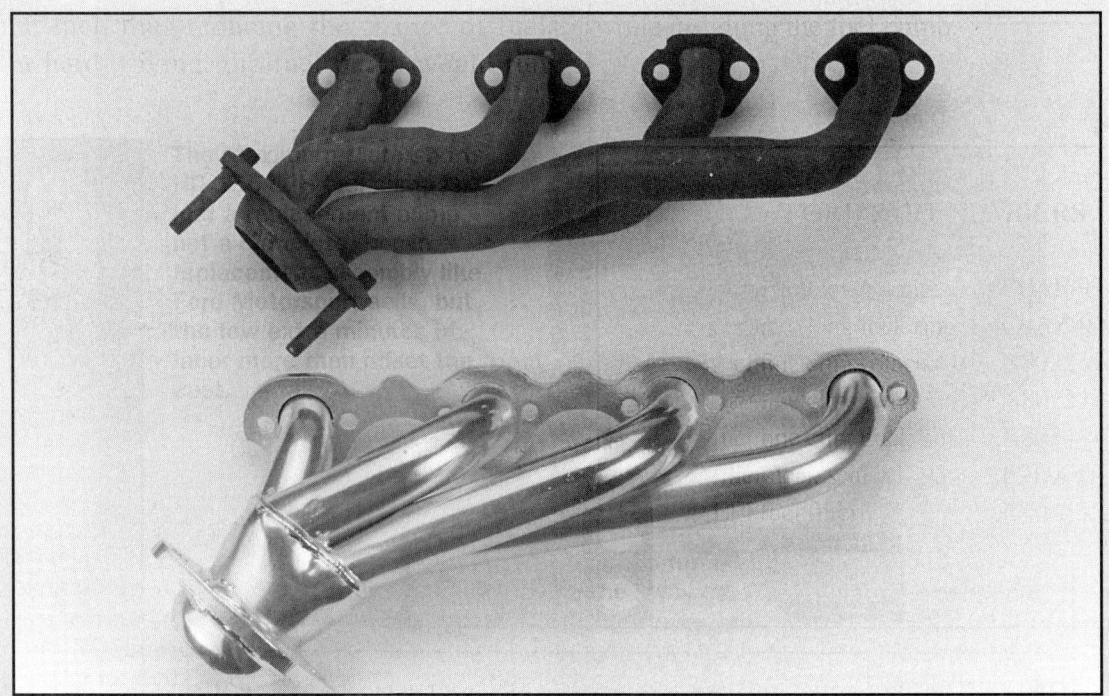

Ceramic coated shorty headers, such as these unequal length 1 5/8" primary tube headers from Walker, will help your 5.0 breathe easier.

SHORTY HEADERS

1. Begin your header removal by disconnecting the HEGO sensors from the catalytic converter H-pipe to access the header to H-pipe mounting flange attachment nuts. Unplug the sensor from the harness and remove the sensor with a 7/8" wrench. Do not drop the sensors, you will damage the ceramic internals. Place them on your workbench out of the way of possible harm.

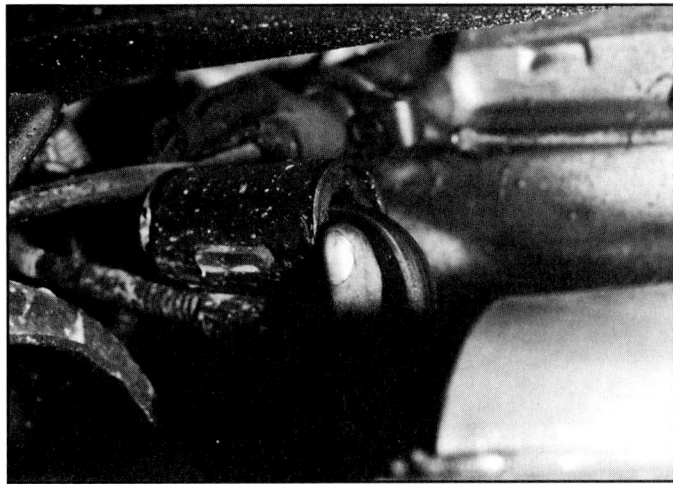

2. Once the HEGO sensors have been removed, it is simply a matter of removing the header flange nuts that retain the converters to the factory headers and prying the converters back slightly until the two flanges are free from the header studs. This hanger shown here is easily used to pry the converters back.

3. Working under the hood now, remove the air inlet tube and Mass Air sensor (if equipped) and unplug all plug wires from their spark plugs. Don't forget to mark them if they aren't already (factory wires are marked with cylinder number). If you are worried about possibly breaking a spark plug during the header removal, you might want to remove your plugs at this time too.

4. At the passenger side rear of the engine, remove the one vacuum line from the Thermactor Air Diverter valve and remove the two hoses that are clamped to the upstream and downstream air tube check valves. Position the two hoses and diverter valve out of the way. Unbolt the downstream air tube from the engine bracket with a small 10mm socket or boxed end wrench and move the tube out of the way.

SHORTY HEADERS

5. Remove both left and right side headers with a 9/16" deep socket. Save the engine lift brackets for later use.

6. Remove the dipstick and dipstick tube from the engine with a gentle twist and pull motion. It will be necessary to grind away part of the forward tab for proper seating next to the new headers. Because of the bigger tubes the dipstick tube will not seat correctly behind the header bolts. Do not leave the dipstick tube loose; it will vibrate loose and either break or fall out.

7. Grind away enough of the inside edge on the right side engine bracket for reinstallation over the header tube. This will be necessary for proper retention of the downstream air tube. If you don't, the tube will eventually crack and break, due to vibration, and cause an exhaust leak. Some header companies actually provide a bracket that does this job for you. Installation of the driver's side bracket is optional. It will be necessary to reuse two factory header bolts to install the OE lift bracket, if that is what you use. The aftermarket header bolts aren't long enough for the header and the bracket together.

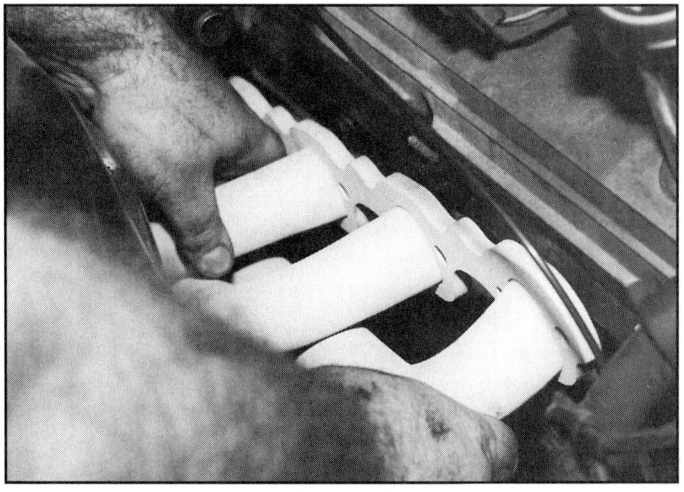

8. Position the headers in place and start the two outboard header bolts. It may be wise to note here that some headers, such as these from Walker's Thrush division, do not use studs, making for a two man operation when it comes time to tighten the headers to the cats. If you wish, you can have the bolts tack welded to the headers to simulate the factory studs. This will make the installation much easier.

SHORTY HEADERS

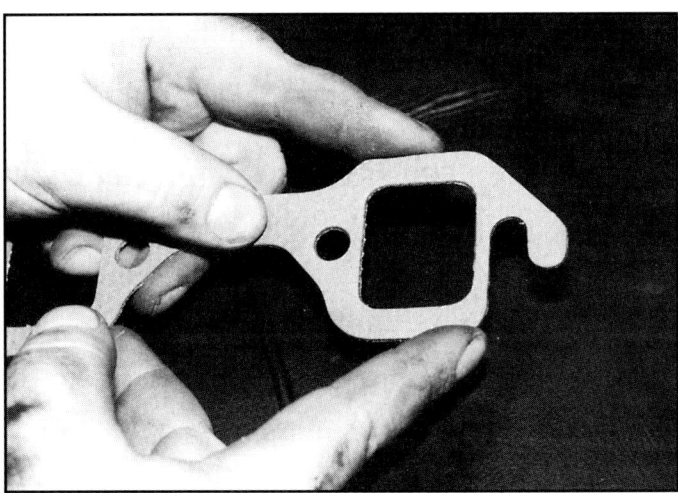

9. We used the header gaskets that came with the headers. Most all header gaskets are made with this notch; if not you can cut the hole open and make a notch out of the outer holes to duplicate this. The reason for the open ends is to allow slipping the gasket between the header and cylinder head and over the two outer bolts you just installed.

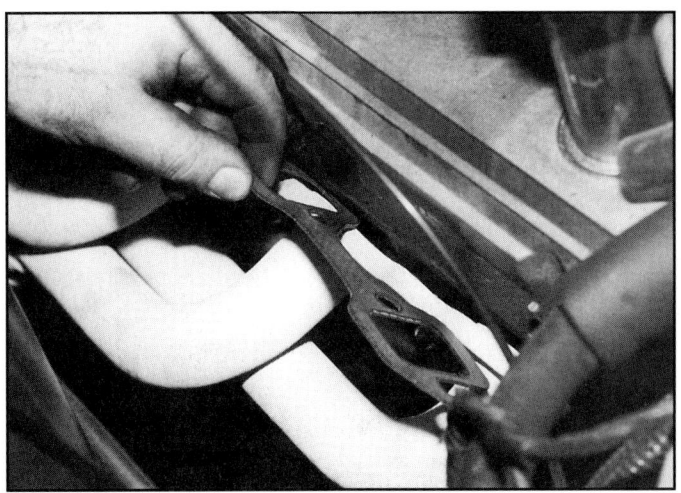

10. Slide the gasket in place between the cylinder head and the header. Once the gasket is in place, install the rest of the header bolts finger tight.

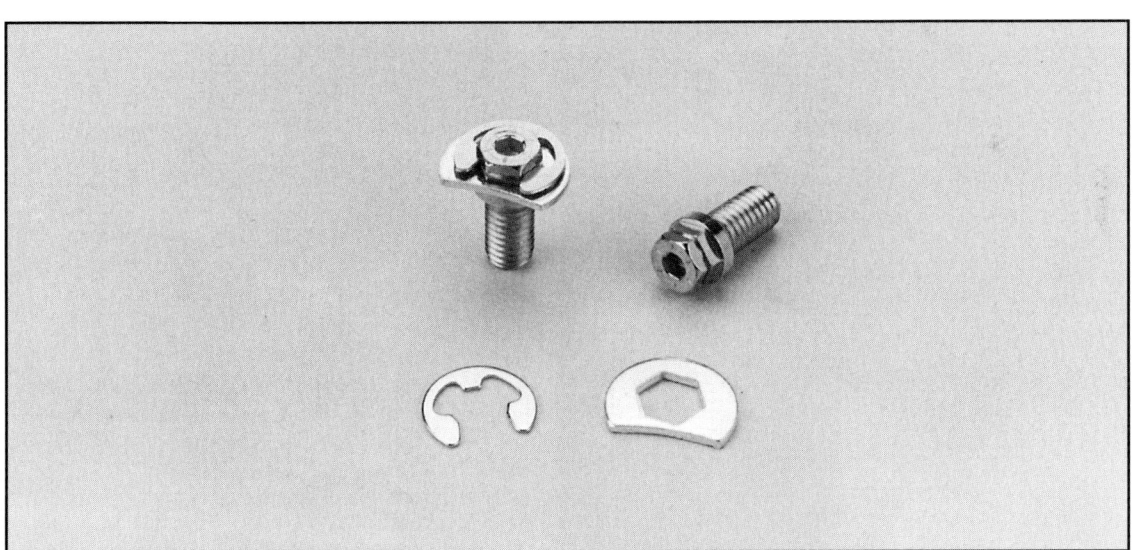

11. We used Stage Eight locking header bolts. These incorporate a locking tab that is secured to the header bolt by an E-ring to prevent loosening of the header bolt, thereby eliminating leaking headers that have to be re-tightened. These are excellent fasteners and have yet to have one single failure. They have further applications for flywheels, intakes, starters, etc. Once all bolts are in finger tight, tighten the bolts and install the locks. Reinstall the converter H-pipe to the header flange and then the HEGO sensors to the H-pipe, start your Mustang and check for exhaust leaks.

MAF METER SWAP

Performance Parts Inc. is the brain child of Jim Dingell. Jim has his hands in quite a few ventures these days, all pertaining to late model Mustangs. Jim heads up the Mustang Special Service Registry, which is a registry for police package Mustangs, in his spare time between handling a real job, and PPI. PPI originated as a mail order business that could offer more for less. With little advertising and overhead, Jim has been able to chop away a few bucks here and there on prices and pass the savings on to the consumer. Every year PPI grows by offering rare, hard to find, special service, SVO, obsolete, and other used Ford parts for less.

PPI is marketing a Mass Air Flow conversion kit, using a variety of new and used parts depending upon your budget. The base kit is a stock MAF powertrain control module, stock '89-'93 5.0 HO MAF meter, and an '89 MAF wiring harness. All items are used and guaranteed to work. Why couldn't you do this yourself you might ask? Actually, you could. But unless you really know a well stocked salvage yard that is willing to let the parts go separately (trust us, this is rare) and one that is willing to warranty or guarantee the electronics, it is much less of a hassle to simply order the PPI kit.

Several options exist, including a Cobra MAF meter, and Cobra PCM's. PPI does request that you call to check for in-stock items first, as items sold as new, such as the Cobra PCM, may be back-ordered, as PPI has to order through special channels for some of these items.

What is the advantage of installing a MAF kit such as this. Well besides the usual reasons you have heard before, such as the ease of installing a larger camshaft, drivability, etc., the biggest reason is savings. How does almost 200 dollars off of the typical MAF swap kit sound when using all used parts? That 200 dollars could go towards your first post-MAF modification.

The PPI MAF basic retrofit kit includes the correct harness, an "A9L" powertrain control module, and stock sized MAF meter. Options include Cobra and aftermarket meters, and the Cobra PCM.

MAF METER SWAP

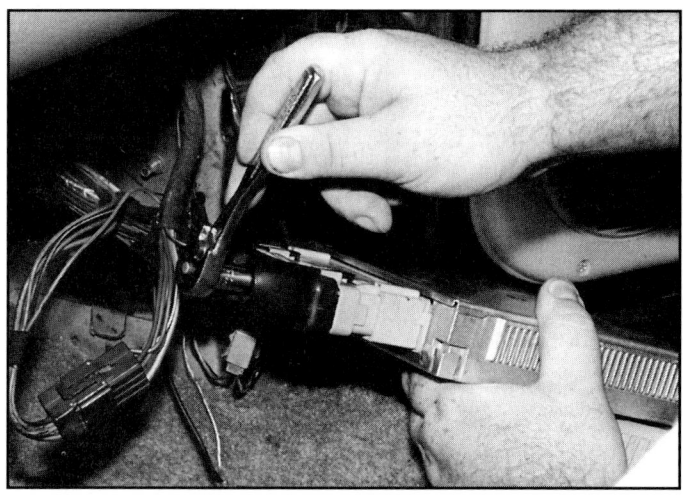

1. Remove the right front kick panel to access the PCM. Once the 7mm bolt is removed the plastic PCM holder can be moved, allowing the PCM to be pulled free of the kick panel area. A 10mm bolt (permanently fastened to the harness connector) retains the PCM to the harness.

2. Once the PCM is disconnected from the main harness, unplug the green eight wire connector, found in the same harness, and unbolt the ground wire under the carpet towards the front of the vehicle.

3. There are two items that need to be removed from the back of the passenger strut tower. The EGR control solenoid (shown here) and the thermactor air bypass and diverter solenoids (TAB/TAD), found below the EGR solenoid. The front of the right strut tower has one connector and the air conditioning WOT relay is located next to it. These must be disconnected/removed from the car at this time too.

4. The five vacuum lines leading to the solenoids: vacuum source, EGR, TAB, TAD, and reservoir all must be disconnected next. You can mark the lines if you wish but the lines pretty much guide themselves to the proper locations.

MAF METER SWAP

5. Once the solenoids are unbolted and out of the way the main harness weatherproof gasket can be carefully pulled free from the firewall and the remains of the harness from inside the car can be pulled through carefully. The EEC-IV relay (the brown one near the PCM 60 pin connector) may need some persuasion from its home high up in the dash.

6. Following the main harness across the firewall you will next come to the two ten pin connectors at the back of the intake, one black and one white. Disconnect both of these. The TFI module pigtail will need to be disconnected from the distributor and fed back to the firewall. Blind hand work under the upper intake will get you there, or you can remove the upper intake if you wish. We were able to blindly pull the harness back and forth for proper removal and installation. We have also unbolted the speed density's MAP sensor at this time. Don't forget to plug the vacuum line coming from the intake, as the BMAP for the MAF system will not need a vacuum signal.

7. There are two eight pin connectors, one black and one gray, in the driver's side corner near the wiper motor that will need to be unplugged from the sub-harness.

8. The final destination of the main harness is the ignition coil and starter solenoid at the driver's strut tower. The four connections that will be tackled here are from left to right: solenoid battery side, ignition coil, negative battery cable PCM ground, and two pin connector near the power steering pump. Once all these connectors are free the harness can be lifted from the car. If you have a strut tower brace, such as our '87 did, you might have to remove it, though we were able to work around our Saleen unit.

MAF METER SWAP

9. Lay the new harness in place and reconnect all electrical terminals. All plugs are the same, so no forcing should be necessary. PPI includes the solenoids, relays, and BMAP from the harness with their kit so nothing is needed to buy. The BMAP shown here has a little cap to prevent the installation of a vacuum line.

10. Our only modification needed was joining these two vacuum lines. The new harness has a long vacuum line to the reservoir in the fender, but instead of removing the wheel and fender liner for access we joined the old line (coming out of the fender opening) to the new line with a "T" (though a straight connector would work if you have one; we did not and the parts store was closed). If you want to remove the wheel and fender liner, by all means do so, but a .50 vacuum line connector will save you a half hour of work.

11. The stock MAF meter comes with all needed ducting and clamps. Simply install it in place of your speed density hose and mount the bracket to the strut tower. Don't forget to plug the MAF meter into the harness.

12. The complete kit looks so stock appearing that most people think our car is an '89. A nice conical filter would be the perfect complement to our new MAF measuring system, now that the PCM can "see" it.

ROLLER ROCKER ARMS

Roller rocker arms have many benefits which include performance increases, lower oil temperatures, and reduced valvetrain and valve guide wear. Aluminum roller rocker arms require less horsepower to operate than their stamped steel factory counterparts and roller rocker arms also increase valvetrain stability. The stock stamped arms are no slouch for any pushrod V-8 engine, but for all-out high rpm usage (where the stamped rockers can flex or even break) the roller rocker arm is the only way to go. Rocker arm assemblies can either have a roller tip, a roller fulcrum, or both a roller tip and fulcrum, depending upon the manufacturer and model type. There are also roller rocker arms with adjustable pushrod cups, and roller rocker arms for pedestal mount or stud mount type cylinder heads. As with most performance parts, you get what you pay for, and the more you spend, the better the rocker arm you will get.

Roller rocker arms have been used, quite successfully too, for several decades in all forms of automobile racing. Their high cost and relatively "exotic" state kept them off of all but the most expensive of super cars and out of the public's reach for many years. But the automakers, always looking for advantages in performance, fuel economy, and noise reduction, began to adapt roller rocker assemblies to their high end and performance lines, and today, many econoboxes and utilitarian vehicles use roller rocker arms as well as rollerized camshafts, lifters, and timing chains for durability, timing accuracy, and longevity. The manufacturers use racing to glean information about parts longevity, aerodynamics, and a multitude of other areas that help them build better cars for you and me.

This is why performance parts that, at one time, were race only parts, can now be found at your local parts store or Ford dealer and installed on your production vehicle.

The rocker arms being installed here are the Ford Motorsport 1.72 ratio for pedestal mount cylinder heads, but Ford Motorsport "Cobra" 1.7s or aftermarket rockers, including stud mount type (which take a few extra steps), can be installed following the same guidelines we are showing here. It would be wise to plan ahead by obtaining a set of valve cover gaskets and a set of rocker pedestal shims. Some roller rocker arm kits include the shims, but if they don't Ford Motorsport carries them under part number M-6529-A302 to maintain proper lifter pre-load. You may not need these shims, depending upon stock head variances, but having a set handy is better than trying to find a set at the last minute with your Mustang all apart.

If you aren't the original owner, or you aren't sure if any other repair or performance work has been done, do yourself a favor and check piston to valve clearance or at least have the cylinder volume checked before modifying your valvetrain. Doing so will ensure you won't have any interference problems with the higher lift the new rocker arms will give your camshaft profile. You also may wish to procure a set of pushrods a few thousands of an inch longer, or at least make sure your local parts house has them in stock, as with some head designs aftermarket rocker arms may need longer pushrods. The rocker arm manufacturer's tech lines should be able to answer any product specific questions you may have about your particular Mustang's setup.

ROLLER ROCKER ARMS

1. The 1.72:1 roller rocker arms are available as a complete set and in sets of two for repair procedures or for mixing ratios on intake and exhaust valves. In the foreground of the photo are the pedal inserts at the left and the Allen head cap bolt kit to the right. These are included with the Ford Motorsport roller rocker arm package.

2. The upper intake must be removed for access to both valve covers. Unplug and tag all electrical and vacuum lines to the upper intake, the throttle linkage, and the valve cover breather hose, and remove the six 13mm retaining bolts (two under the engine ID intake plate) to remove your upper intake. Some engines may have a dampening bracket bolted between the right rear upper intake bolt and the left rear lower intake bolt. Inspect for this bracket before trying to pry away your upper intake to prevent intake manifold damage. The throttle body inlet hose has also been removed in this photo.

3. You may wish to use a short length of metal fuel line to connect the two EGR cooler lines together to prevent coolant spillage during upper intake removal. Another smooth tip is to install two vacuum caps or a length of vacuum line to enclose the two EGR cooler nipples. This will prevent any stored coolant in the EGR spacer from pouring out when removing the upper intake.

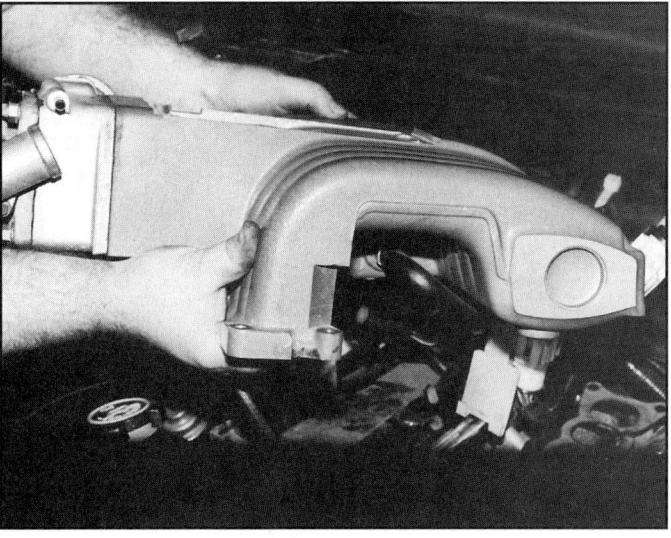

4. Carefully remove the upper intake from the engine. If you have a strut tower brace you might have to remove it, or you may be able to work the intake off around it. With the various different manufacturers of strut braces, there is no way a definite answer can be given. You will simply have to decide for yourself if the brace needs to be removed.

ROLLER ROCKER ARMS

5. Place a length of duct tape securely over all eight intake holes to prevent any contamination of the cylinders by gasket particles or stray bolts and washers. This could be devastating to the engine upon start up and should be masked off immediately after removing the upper intake and before ANY other bolts are removed.

6. Remove all four plug wire retainers from the valve cover bolt studs (two per side) and then proceed to remove all six bolts (per cover) with a deep 7/16" socket and gently pry the covers up and off of the engine. Place the valve covers in a cleaning solvent to soak while working on the rocker arms.

7. Remove all major sections of the old valve cover gasket and then, using a gasket scraper, remove the remaining pieces. Do not let any gasket chips enter the oil feedback holes in the corners of the head.

8. Remove all 16 stamped rocker arms, fulcrums, and retaining bolts from the cylinder heads. The fulcrum guide usually sticks to the fulcrum because of a light film of oil between them. The fulcrum guides will be reused with the roller rockers (Cobra 1.7 rockers come with new guides). When removing the stamped rockers ensure that the pushrod remains in the bucket of the lifter.

ROLLER ROCKER ARMS

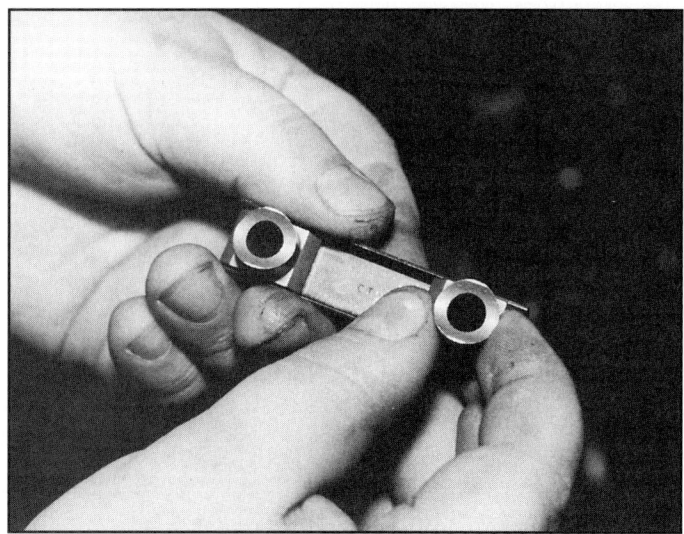

9. Insert two fulcrum adapters from the roller rocker arm kit into the fulcrum guide and place the assembly onto the cylinder head pedestal.

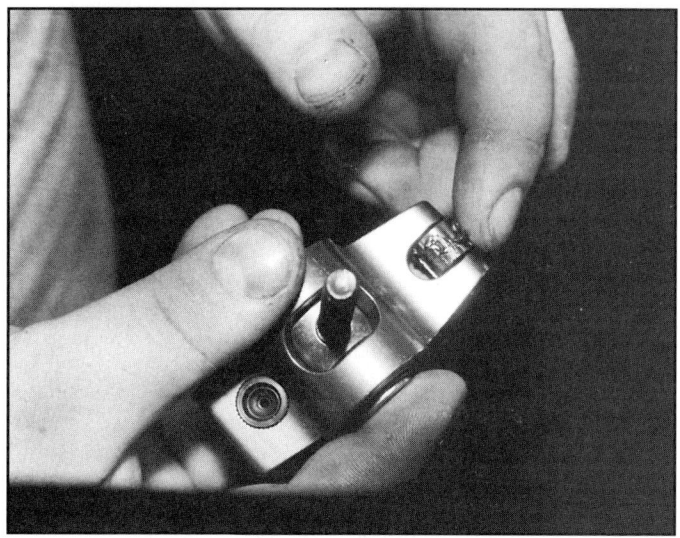

10. Lightly grease the roller tip and the pushrod bucket with a moly grease, such as what is used with new camshaft installations.

11. Insert an Allen head cap bolt through the roller rocker fulcrum (ensure that the flat side of the fulcrum is facing up to receive the screw head) and tighten the cap screw snugly, but do not torque the cap bolts at this time. Ensure that the roller tip does not move off of the valve stem tip while tightening the cap screws. Repeat this for the complete valve train system.

ROLLER ROCKER ARMS

13. We used a spray on gasket adhesive to secure the Mr. Gasket Ultra Seal valve cover gaskets to the cylinder head. You may wish to adhere the gasket to the valve cover itself instead of the cylinder head. Either way is acceptable as long as the bolt holes are lined up and the gasket is properly positioned between the two sealing surfaces for proper oil retention.

12. Once all 16 roller rocker arms are installed to "snug", turn the engine over, clockwise to top dead center (the timing pointer pointing at "0" on the crankshaft), and check the number one cylinder intake and exhaust valves. They should both be closed (you should be able to spin the pushrod). If they are not both closed the crankshaft will need to be rotated 360° to TDC to correctly start the torqueing procedure. With the engine on TDC, torque the number one intake, number one exhaust, number four intake, number three exhaust, number eight intake, and number seven exhaust rocker arms to 15 lb.-ft. with a torque wrench. Once torqued, the pushrods should still be able to be rotated. If the pushrods are tight and will not rotate, the shims will have to be implemented. Begin with the thinnest shim and repeat the above procedure until the pushrod can be rotated without any end play. Once these valves have been checked and the rocker arms adjusted, continue to the next set by rotating the crankshaft 360° and torqueing the number three intake, number two exhaust, number seven intake and number six exhaust rocker arms in the same manner as the previous set. Complete the torqueing procedure by rotating the crankshaft an additional 90° and torque the number two intake, number four exhaust, number five intake, number five exhaust, number six intake, and number eight exhaust rocker arms as above. Most engines will not need the shims, but it is better to be safe than sorry.

14. Remove the three small bolts from the passenger side valve cover oil baffle. Discard the oil baffle and install both valve covers. You will need to remove this oil baffle for roller rocker arm clearance. The Ford Motorsport chrome plated HO valve covers will also fit over these roller rocker arms, but factory non HO stamped steel covers will not work. Inspect for proper clearance and grind away any interfering parts of the inside of the valve cover for proper clearances.

ROLLER ROCKER ARMS

15. Reinstall the valve covers, ensuring a proper seal. Start all bolts evenly and then tighten to 10-13 lb.-ft.

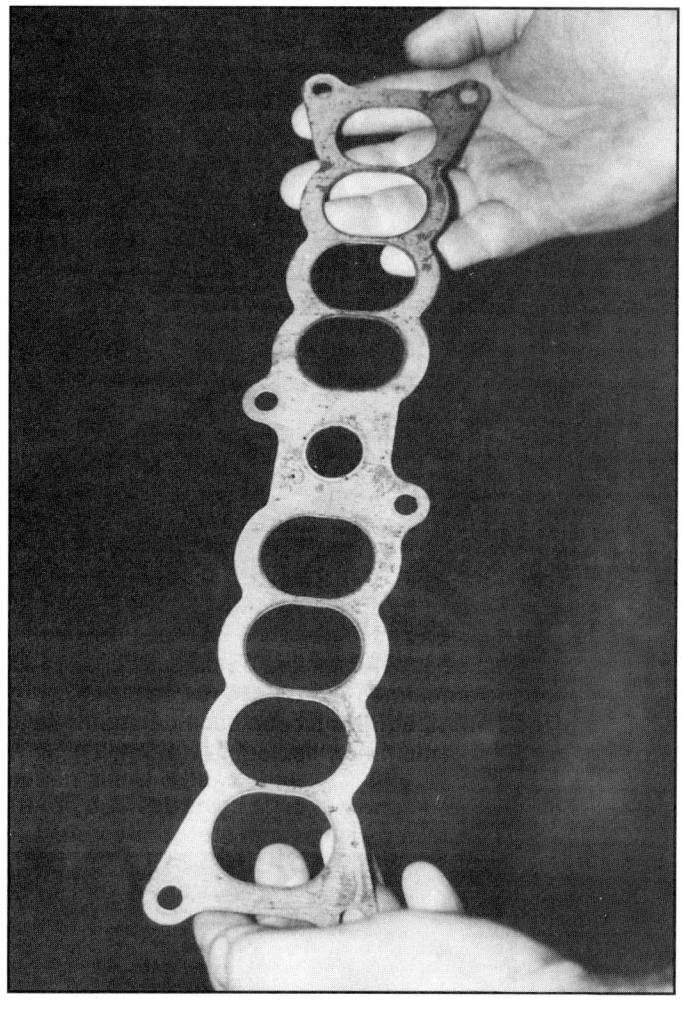

16. Inspect the upper intake gasket for any signs of sealing damage. Generally speaking, with careful removal procedures the upper intake gasket can be reused three to five times. If there are signs of damage or you simply feel safer with a new gasket, obtain a new upper intake gasket either from Ford (E6SZ-9H486-C) or from Mr. Gasket (#148). Reinstall the upper intake manifold and torque the retaining bolts to 12-18 lb.-ft. Reconnect all electrical and vacuum connections previously tagged and removed. Start your vehicle and listen for any excessive noise. There should be no noticeable increase in valvetrain noise.

LOW TEMP THERMOSTAT

Not many people think that a thermostat can make more power for your Mustang. The answer to that is yes, but only when using a quality high performance thermostat. The most common type of thermostat change for the 5.0 crowd is to remove the factory 192° unit and install a low temperature 180° or 160° unit. While you may feel that this is going to lower your coolant temperature by opening the thermostat sooner, in actuality you will see very negligible performance gains, if any at all. You may not see any cooling benefits either when you purchase a two dollar 180° replacement thermostat at your local parts store. The reason behind this is the fact that with conventional thermostats, the high coolant pressure actually alters the opening temperature point drastically. For example, at 30 pounds of coolant pressure, a 185° thermostat will actually not begin to open until 198.5°. That, my friend, is a whopping 13.5° difference. Not very cool now is it, pun intended.

In steps Mr. Gasket, makers of high performance parts for many years. One of their newest parts to be added to their late model performance line-up is their complete selection of balanced high performance thermostats. With a Mr. Gasket 180° balanced thermostat placed in the same condition as just described, it will have altered its opening pressure by only 2° at 30 pounds of coolant pressure. What exactly is a balanced thermostat and why is it different than the dime store varieties, you say? Let's start with their design. The large three port opening of the Mr. Gasket unit equalizes coolant pressure above and below the thermostat's moving valve. This equalization pushes against the thermostat's sleeve equally from either side and "balances" it, letting the thermostat respond to coolant temperature changes only. The thermostat is designed to open at the correct temperature, regardless of rpm, and responds quickly to coolant temperature variations, effectively reducing the effects of wide coolant pressure swings, which can cause the valve to stay closed longer than it's supposed to and causing an overheating condition. The design of the sleeve and the ports allow for self cleaning and increased coolant flow to the radiator, making for a superior thermostat assembly which will do what you wanted to do in the first place, make power and keep your car cool. A half hour installation on a cold engine and ten dollars out of your wallet will improve your Mustang's operating conditions immensely.

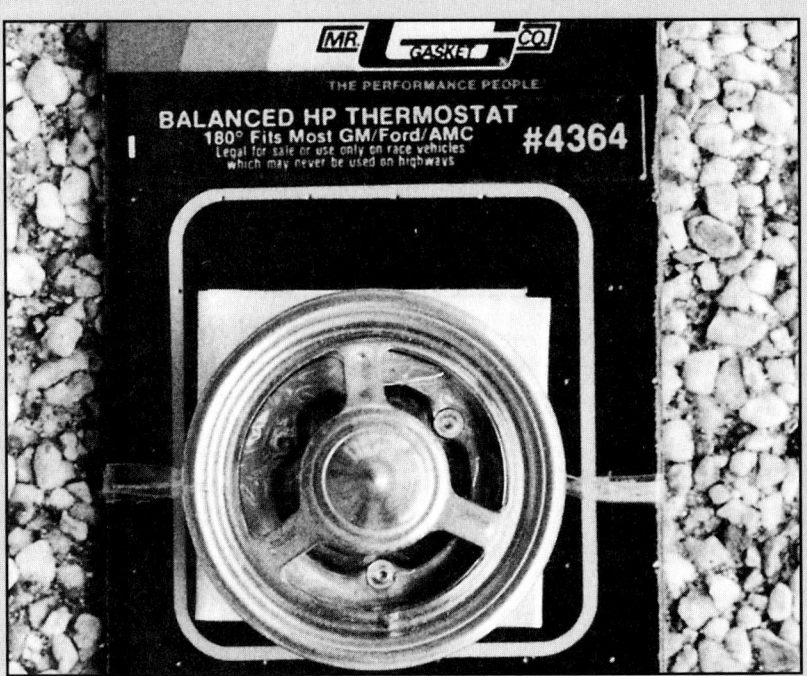

The High Performance Thermostat from Mr. Gasket is available in 160°, 180°, and 195° increments. We opted for the 180° unit (part # 4364) instead of the 160° unit (part # 4363) for better driveability. The 160° unit might not allow the Powertrain Control Module to obtain open loop control. You will also need a new thermostat housing gasket (part # 746) from Mr. Gasket for the replacement procedures.

LOW TEMP THERMOSTAT

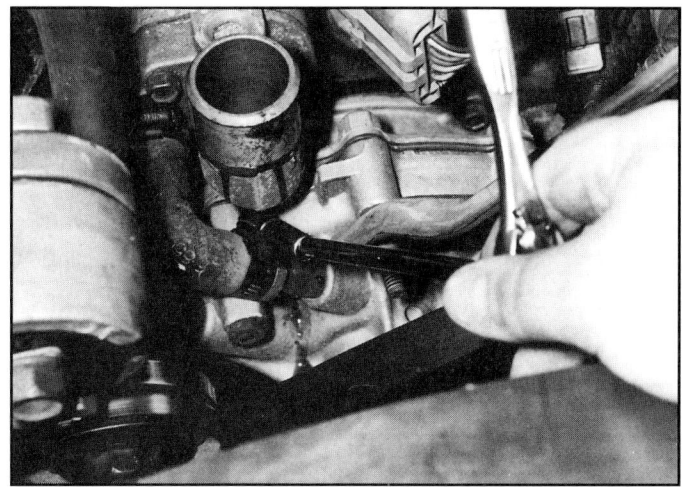

1. After draining your coolant into a clean container (you might want to consider a cooling system flush at this time) from the radiator drain petcock, remove your upper radiator hose from the thermostat housing. If you have a spring loaded clamp (as seen here) on your hose, do not reuse it. Replace it with a worm drive clamp.

2. Loosen the two small worm drive clamps at either end of your coolant bypass hose and remove the hose from your vehicle. Inspect your hoses at this time and replace as needed.

3. The distributor will have to be moved out of the way to gain access to the housing's lower bolt. Mark the location of the TFI ignition module in relation to the a/c bracket with a grease pencil or crayon to properly orient the distributor during reassembly. It would be advisable to place a piece of cellophane tape over the grease mark, as to not wipe it away accidentally.

LOW TEMP THERMOSTAT

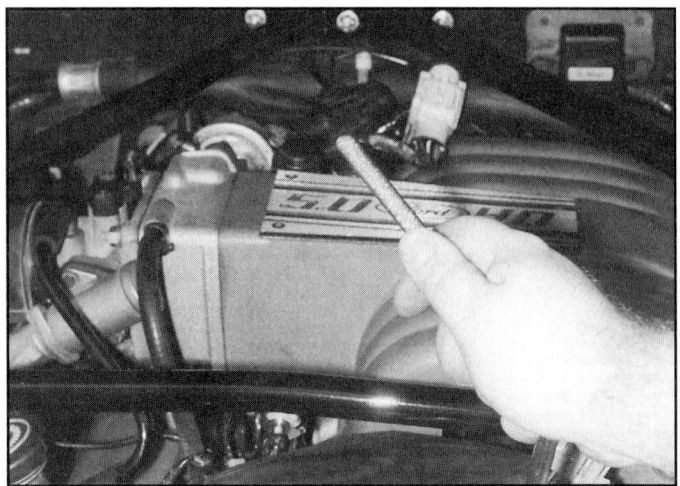

4. Using a 1/2" distributor wrench (a swivel socket might get you by), loosen the distributor hold-down bolt roughly two turns. You don't need to remove the hold-down assembly, just have it loose enough to turn the distributor counter-clockwise and gain the needed access.

5. Loosen and remove the housing retaining bolts from the lower intake assembly.

6. Lightly tap the thermostat housing to break it free from the lower intake. The lower retaining bolt will come out with the housing, as it cannot be removed completely due to the timing cover.

LOW TEMP THERMOSTAT

7. Clean both the lower intake and housing mating surfaces of all gasket material.

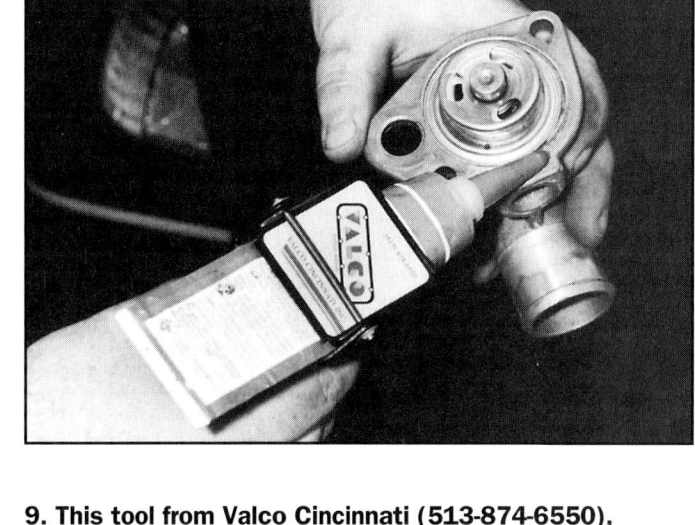

9. This tool from Valco Cincinnati (513-874-6550), called the Tube-Grip, allows precise applications of silicone adhesives and other tube dispensed shop items. No longer will you have to smear silicone into tight areas with your fingers. Possibly the greatest thing about this trigger type device is that when you stop squeezing the handle, the silicone stops flowing! Try that one with hand squeezing a tube sometime. We used a thin application of adhesive to the housing to retain the gasket while reinstalling the thermostat and housing assembly. It is critical to install the thermostat correctly by having the cone facing the radiator. Another critical factor is ensuring the gasket will retain the thermostat during installation. If the thermostat slips out of its mounting groove, the gasket will not seal the housing to the intake. Once the thermostat is installed, reposition the distributor to your marking and tighten the mounting bolt. Close the radiator drain petcock and refill the radiator. Once hot, check for any leaks and top off the radiator and coolant recovery bottle while the engine is running to remove any air pockets.

8. This comparison of the factory 192° and the Mr. Gasket 180° thermostat shows the differences in the flow areas.

THROTTLE BODY & EGR

In today's emissions legal parts arena there are some budget priced parts that not only enhance your Mustang's power the second you're done turning wrenches, but can enhance further modifications down the road. For example, shorty headers and free flowing mufflers not only allow your car to breathe better when you bolt them on, but will allow your 5.0 to make more power when you add rocker arms or camshafts by expelling the gasses all that more efficiently.

The 5.0 Mustang's throttle body is one of many areas that can be improved by the addition of an aftermarket unit. While bigger is not always better, as in the case of header or exhaust pipe size, the throttle body size should be conservatively chosen. While the stock 58mm throttle body is suitable for production engines, an increase in throttle plate size will allow for an increase in incoming air charge, which results in a horsepower increase when the Powertrain Control Module adds the necessary fuel at the injectors.

For years the most common size throttle body to upgrade to has been the 65mm Ford Motorsport unit. But within the past few years, as the 5.0 race crowd got faster and faster, there was a need for larger and larger throttle bodies. While today there are throttle bodies in the 80-90mm size (large enough to pass your hand through), most of these are manufactured from billet aluminum. While these race-only throttle bodies are not for the street, there was a need for an emissions legal alternative to the power restricting stock unit.

BBK Performance has been in the 5.0 Mustang parts business from practically day one, and with their vast resources have come up with a selection of throttle bodies in the 65, 70, and 75mm sizes, with matching EGR spacers that are all CARB approved for your street driven Mustang. BBK also offers their own 80mm race version throttle body, which is not street legal. Once we obtained our 70mm throttle body and EGR spacer from BBK it would take about an hour to turn the wrenches and hit the local 1320 for some fun.

The BBK 70mm throttle body as the incoming air charge would get a quick glance at it. We complemented the 70mm throttle body with BBK's matching 70mm EGR spacer.

THROTTLE BODY & EGR

1. Make sure the car is cool before disconnecting and removing all components associated with the stock throttle body. Here the air bypass valve has already been removed and the throttle linkage is being disconnected. The linkages simply unsnap from the throttle body linkage arm.

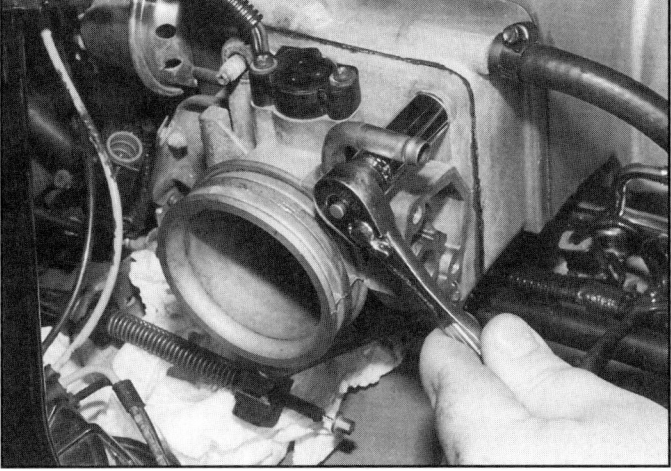

2. Once everything has been disconnected from the throttle body, the four 13mm retaining nuts can be removed from the upper intake studs.

3. With the four nuts removed, the throttle body can be separated from the EGR spacer. A towel or rag should be placed under the throttle body during the separation to catch the small amount of coolant that is trapped in the EGR spacer.

THROTTLE BODY & EGR

4. Remove the two 10mm bolts that attach the throttle linkage to the EGR spacer and pull the linkage away from the EGR spacer for added work space.

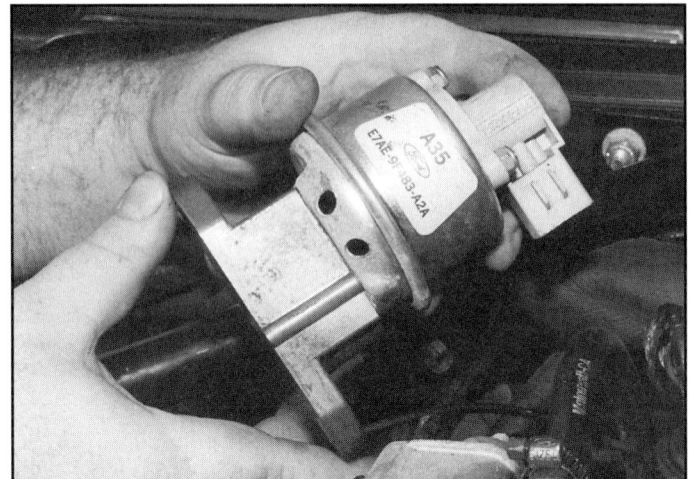

5. The EGR valve will need to be removed from the stock EGR spacer and transferred to the BBK unit. If your car doesn't have a strut tower brace you more than likely will be able to remove the EGR spacer and EGR valve as an assembly and transfer them on the bench. Our 5.0 had a Kenny Brown strut brace installed, which necessitated removal of the EGR valve from the spacer while still under the hood. The EGR valve is attached with two different size nuts. The upper is a 1/2" and the lower is a 9/16". Careful removal will allow the reuse of the gasket.

6. Disconnect the EGR cooler hoses and cap or connect them with a short length of tubing to prevent coolant spillage and remove the EGR spacer from the upper manifold studs.

THROTTLE BODY & EGR

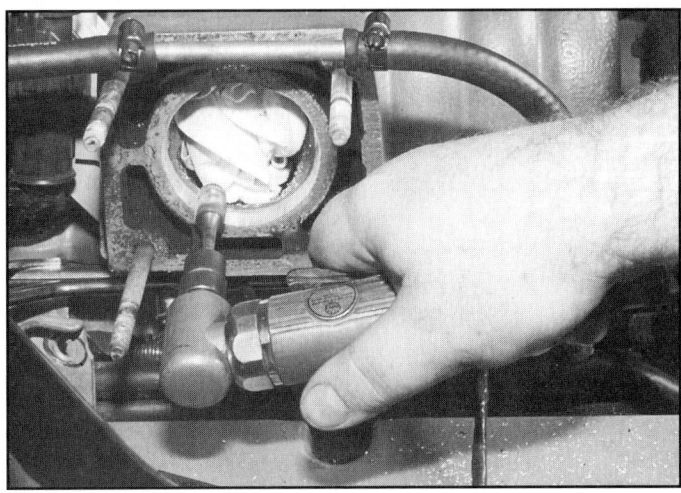

7. BBK suggests opening the upper manifold to the size of the new throttle body, in this case, 70mm. Block off the intake with a shop rag and scribe the new opening, using the new larger gasket or the EGR spacer as a guide. Open the upper manifold to this line with a grinding stone or carbide burr. Vacuum all metal shavings from the mouth of the upper intake before removing the shop rag. Wear mouth and eye protection during this operation; the carbide burr makes very small metal particles which can cause eye and throat pain.

8. In the case of throttle body and EGR spacer installation, when we say installation is the reverse of removal, we couldn't be more correct. Reconnect everything but the air bypass valve connector. Set the idle speed and set the throttle position sensor as per BBK's instructions.

9. Once done with throttle position and idle settings, reinstall the electrical connector to the air bypass valve. One thing we did note with our project car was our AOD seemed to lose some of its shift firmness after installing the BBK units. A TV (throttle valve) adjustment brought our AOD's shift characteristics back up to snuff in no time. To make sure there is no slack in the TV cable at the throttle body, simply disengage the small white locking clip on the TV cable and push the TV cable head against the TV spring (see photo number 1) until there is no slack left in the spring. We added about an 1/8 of an inch more tension once the spring's slack was removed to add even more shift firmness to our AOD transmission. A quick call to BBK assured us that this was not a common problem and that we might have bumped the cable during removal. Which is possible, we are human.

K&N FUEL INJECTION PERFORMANCE KIT

Attention all 5.0 Mustang owners with conical filters. Do you realize that, unless you have a FIPK system on your Mustang, your filter isn't CARB legal. If you have underhood inspection programs in your area that conical filter will stick out like a big red flag. These and other reasons are why K&N developed the FIPK system. FIPK stands for Fuel Injection Performance Kit. The FIPK removes the stock filter housing and filter and replaces it with a conical high flow reusable K&N FilterCharger, and all necessary brackets and tubes to retain the emissions legal status needed for 50 state approval. The FIPK system is available for all EFI 5.0 Mustangs from '87-'95, and '96 FIPKs are in the works. The '94-'95 FIPK is the newest one available so we decided to show that installation on a '95 GT first, and follow the '95 installation with a recap on the older FIPK systems.

The FIPK for the '94-'95 Mustang ditches all that plastic filter and MAF meter housing garbage for an honest to goodness open element conical filter, and it's CARB legal too! By removing the excess plastic you will now have access to your MAF meter for adjustments or swapping with relative ease, not to mention the sheer amount of increased air flow that this baby can handle. About the only bad thing we can think of is taking a drill to the inner fender for a mounting bracket. But since the new FOX4 platform doesn't use a MAF meter bracket like the earlier cars, the hole is a necessary evil.

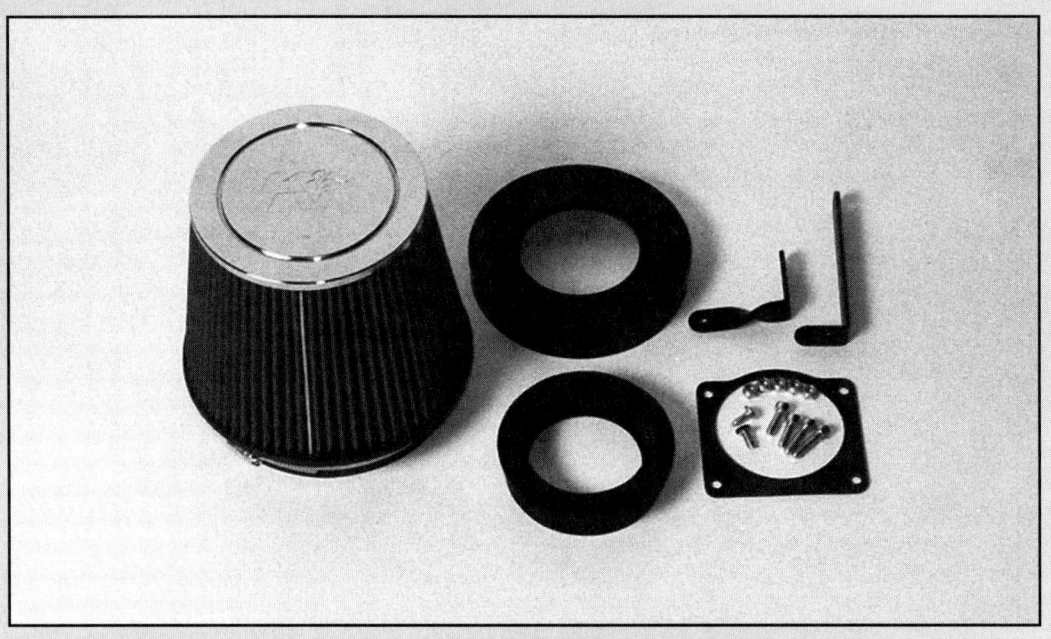

The FIPK system comes complete with everything you will need to deep six your old filter housing.

K&N FUEL INJECTION PERFORMANCE KIT

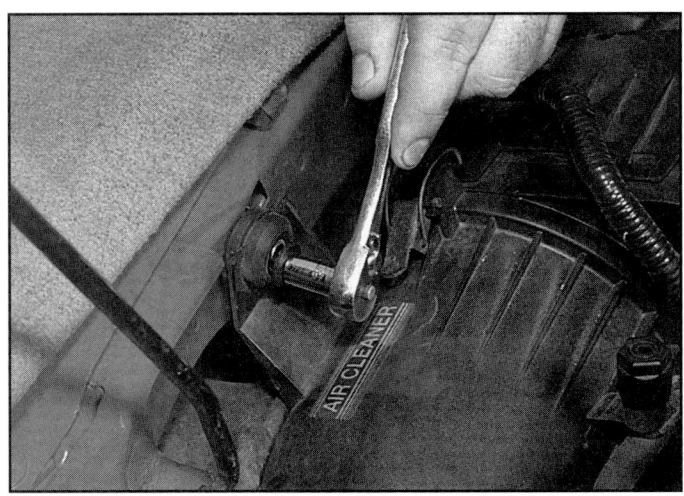

1. This metric retaining bolt will need to be removed. It is not needed in the FIPK system, so it can be reinstalled into the J-clip after the housing is removed or it can be discarded.

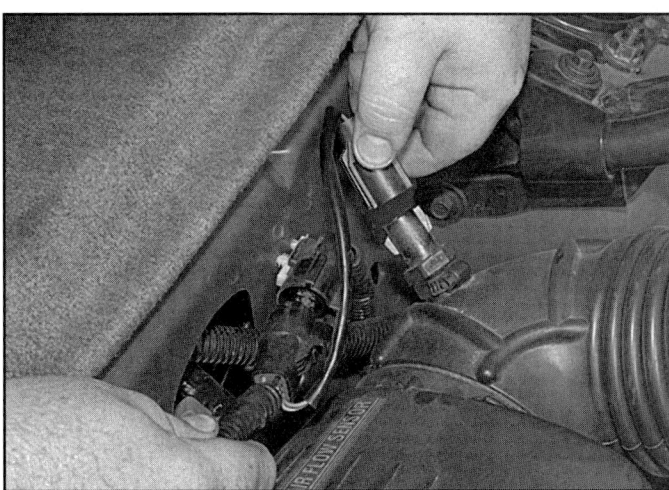

2. Unplug the ACT (Air Charge Temperature) sensor at the inlet elbow and the connector at the fender; this will allow the harness to be removed with the housing. For all you drag racers; put an ice pack on the ACT sensor while at the track, this will fool the sensor into thinking the incoming air is cooler (read: denser) and will make more power.

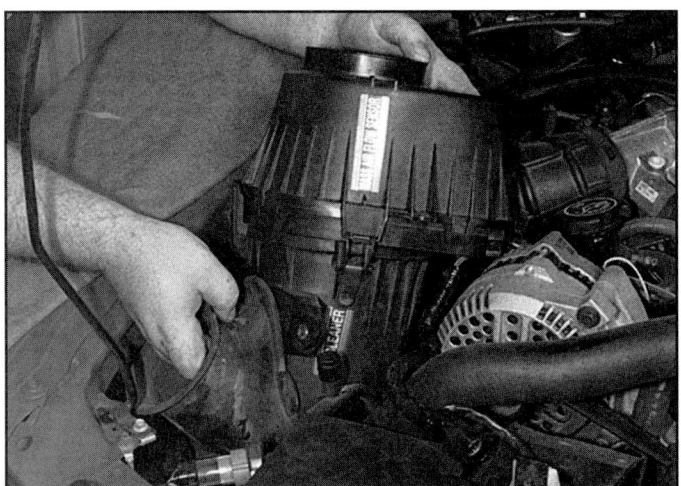

3. Loosen the large worm drive clamp at the MAF housing to air inlet elbow connection and remove the MAF meter/air filter housing from the inner fender. The housing is retained to the fender by a rubber washer in the lower inner fender. Some effort might be needed to pull it free.

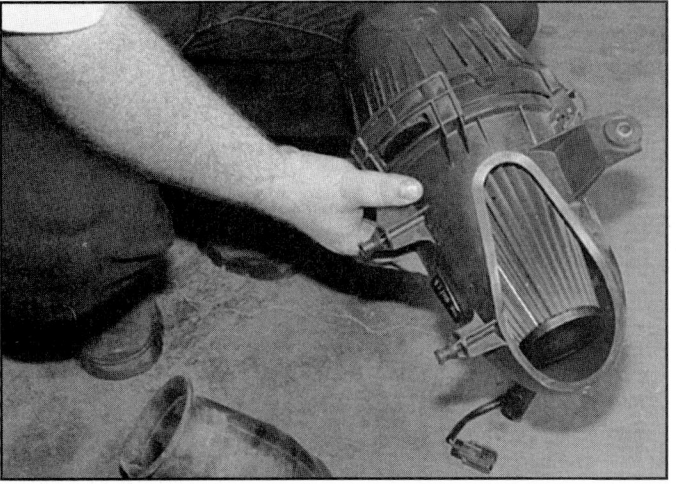

4. Once the filter is free, we removed the rubber mask that directs air from the inner fender area into the filter to show you just how little of the filter gets a "ram" effect. True, the air can circulate around the filter, but there isn't enough room, especially with the stock paper element. Remember, the straighter you get the air, the more air flow you will have.

K&N FUEL INJECTION PERFORMANCE KIT

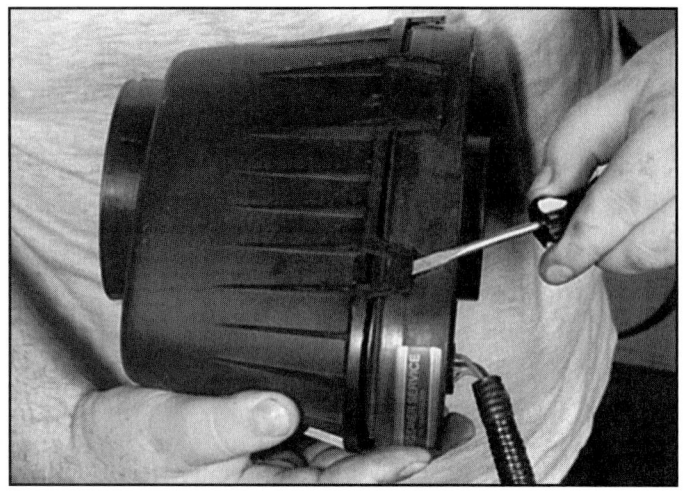

5. After separating the air filter from the MAF housing, use a flat blade screwdriver to pry the tabs back gently that hold the MAF meter mounting plate to the MAF meter housing and separate the two.

6. Disconnect the wiring from the MAF meter and remove the harness from the housing. Set the harness aside for use later. Remove the four mounting nuts that hold the MAF meter to the mounting plate. Set aside the plate and the four mounting nuts, as they will not be reused.

7. Install the stock MAF meter onto the metal filter ring with the supplied rubber seal sandwiched in-between them. There are four metric Allen bolts, two of which are slightly longer and are used to attach the two fender brackets. Position the brackets according to the instructions and tighten the Allen bolts.

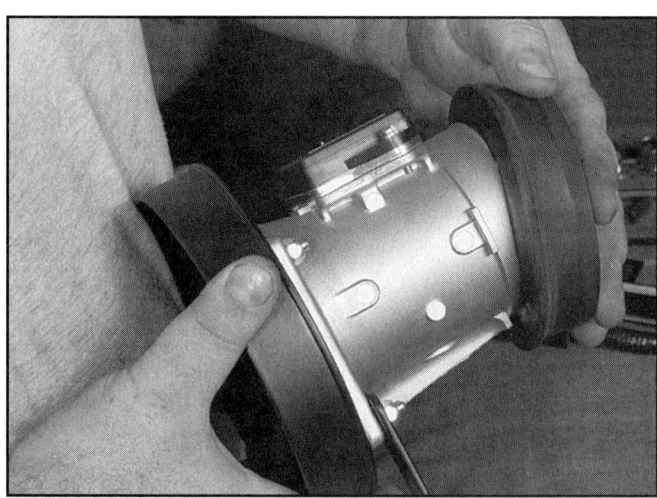

8. Slip the adapter/sealing ring over the outlet end of the MAF meter. The meter is now ready for installation. Set it aside while you finish the under-hood steps next.

K&N FUEL INJECTION PERFORMANCE KIT

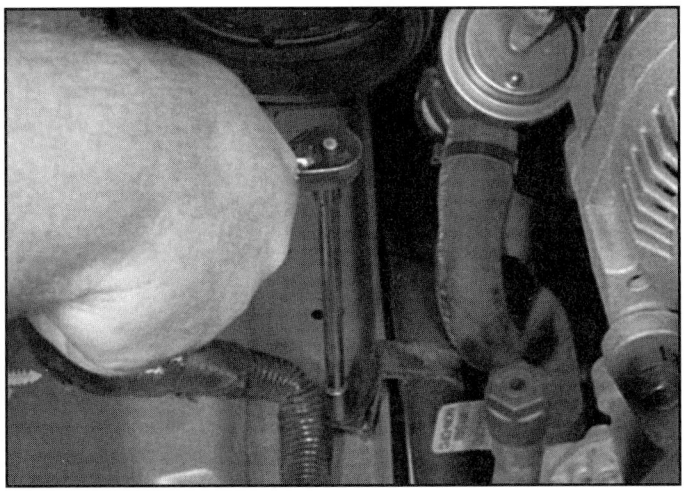

9. This A/C retaining bracket is where the lower bracket will be attached. Remove the stock self tapping screw and set it with your other non-reusable parts; there is a longer self tapping screw included with the FIPK.

10. Here is where the fun comes in. Set the MAF meter and adapter in place. Once everything is secured, mark and center punch for the hole to be drilled in the inner fender. Remove the MAF meter and drill the hole. A right angle drill would be helpful, but you can sneak in a regular drill with caution.

11. Once the hole is complete, touch up the opening with a dab of touch up paint to prevent rust and lower the MAF/air filter assembly into place. Install the new self tapping screw through the lower bracket and the bolt and lock nut into the inner fender bracket. Tighten both hose clamps (at the filter and the inlet hose), reconnect the MAF and ACT wiring and affix the CARB label in a conspicuous place, such as the inner fender above the filter.

K&N FUEL INJECTION PERFORMANCE KIT

The K&N FIPK is also available for '87-'88 non-MAF Mustangs and '89-'93 MAF equipped Mustangs. Here is a quick overview of their installation.

1. Removal of the stock air box, via the two top mounting nuts, is necessary to install the new K&N FIPK. If you haven't removed your air intake silencer yet now is the time to do it.

2. The tubular chrome bracket in the non-MAF meter kit is placed between the stock air inlet hose and the new K&N filter. This acts as the rear support for the new open conical filter. Once you slide the bracket into the stock inlet hose you can align the mounting hole in the inner fender with the bracket fairly easily. The instructions note that the hole may need to be enlarged, but not so on our '87 GT convertible.

3. We loosely mounted the "L" bracket to the face of the filter and then slipped the filter into place for a test fit. The opening of the filter is angled for proper filter orientation. Some slight rotation of the filter will get the "L" bracket lined up and sitting flush with the inner fender. Our forward mounting hole did need a slight cleaning with a reamer for the bolt to slide through freely. If you don't have a reaming tool a large self threading screw can enlarge the hole in a pinch too.

4. Once the hole had been enlarged we reinstalled our filter assembly and tightened down the forward "L" bracket support.

5. Center the inlet hose and filter on the rear support and tighten both clamps, the original one and the new one on the filter. Finish off the installation by applying the CARB certification decal.

IV.
FUEL SYSTEM & INDUCTION TECH

FUEL PRESSURE REGULATOR
HIGH VOLUME FUEL PUMP
NITROUS INSTALLATION
SUPERCHARGER INSTALLATION

FUEL PRESSURE REGULATOR

With the addition of bolt-on performance parts, the Mustang needs extra fuel delivery to handle the extra power increase. Fuel delivery can be increased one of two ways, either by increasing pressure or increasing injector size, both of which require more volume. While a fuel pump is a necessary addition to any modified Mustang, this extra volume needs to be controlled. The stock 19 psi injectors, while good for up to 286 horsepower (see chart on page 75), usually give way to 24 lb. injectors (increasing fuel delivery). With 24 lb. injectors and a 155 LPH fuel pump the Mustang is good for up to 360 horsepower. To tie these components together and make a completely tunable package, an adjustable fuel pressure regulator should be used. While the regulator alone on a 19 lb. injector equipped 5.0 will increase delivery with the higher pressure, you must first make sure you have enough fuel pump to handle the increased pressure (see chart).

Some tuners with EEC-IV engine controls experience have said that a Mass Air Flow-equipped Mustang can actually lose power with the installation of fuel pressure regulator. But what we have found is that on these MAF equipped Mustangs the MAF meter will trim the adjustment and not show a power gain. Matter of fact, a MAF equipped Mustang that is in near stock condition, performance wise, will actually benefit from a lower than stock fuel pressure rating. Speed Density equipped Mustangs will benefit even more from the use of an adjustable fuel pressure regulator.

We obtained Kenne Bell's all billet adjustable fuel pressure regulator, which is fully adjustable to 100 psi and is fully rebuildable. We really like the Kenne Bell regulator as it is a beefy, fully CNC designed regulator, and not some modified stock regulator forced into duty in a performance application. The regulator design incorporates a vacuum signal port for vacuum compensation and a 1/8-inch pipe thread port for use as an auxiliary regulated fuel source. This port can be used for a permanently mounted fuel gauge or to feed an auxiliary system such as a nitrous fuel solenoid.

While this installation can be done in a Saturday afternoon, tuning the car for optimum performance with your particular engine package will take a few drag strip tests to find exactly how much pressure your car likes. Kenne Bell also carries a hood-mounted fuel pressure gauge, if you do not wish to use an under hood unit mounted on the fuel rail.

The billet aluminum adjustable fuel press-ure regulator from Kenne Bell is fully anodized. Notice the Allen head screws around its mid-section. The regulator is fully re-buildable and also has a provision for a fuel gauge or nitrous fuel feed line.

FUEL PRESSURE REGULATOR

1. Remove all hoses and linkages from the upper intake. This includes the throttle and AOD transmission TV cables, the three vacuum lines at the rear of the intake, purge solenoid hose behind the distributor, and all electrical wiring plugs such as the EVP and TP sensors.

2. After the upper intake cover has been removed the two center intake bolts can be removed.

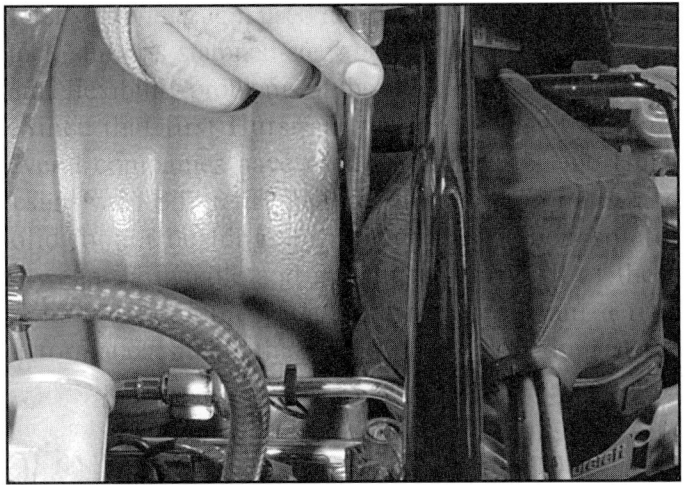

3. Removal of the two front and two rear bolts are all that remain to free the upper intake from the lower. If your upper intake has never been removed before there might be a bracket from the left rear upper intake bolt to the left rear lower intake bolt. This bracket must be removed first and for all intents and purposes can be discarded.

4. Just before removing the upper intake assembly, cover the two EGR cooler ports with a length of hose or two rubber caps to prevent coolant spillage. The two coolant hoses can be mated together with a short length of fuel or brake line to prevent coolant spillage too.

FUEL PRESSURE REGULATOR

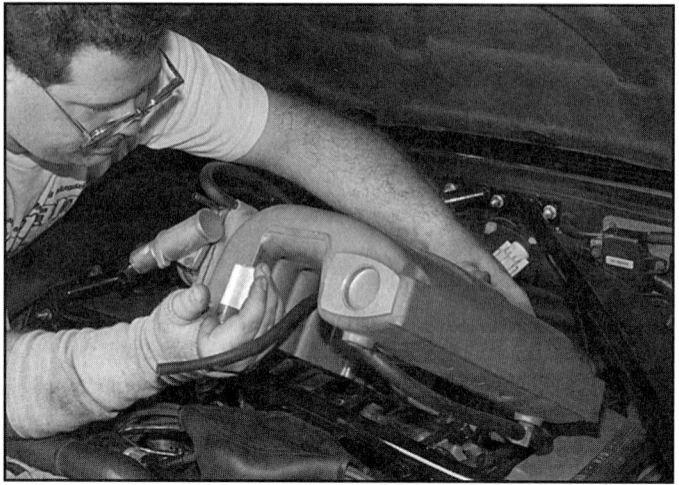

5. The upper intake is now ready for removal. While our Kenny Brown strut brace allowed us to remove our intake without a problem, your strut brace may need to be removed if found in the way of intake removal. As soon as the intake has been removed, cover the lower intake openings with duct tape or shop rags to prevent contamination or engine damage from a stray nut or bolt.

6. With the upper intake removed, the factory fixed fuel pressure regulator is plainly visible. Remove the three small Allen head screws to remove the regulator from the fuel rail. If you feel compelled and have the correct tools, the fuel rail can be disconnected and unbolted from the lower intake for transfer to a work bench to have the regulator removed.

7. The Kenne Bell billet regulator uses the factory Allen screws for installation. Ensure that the new regulator is completely seated to the fuel rail and that all three Allen screws are tight. The fuel pump will run with the upper intake removed; simply turn the ignition key off and on a few times, but do not try to start the car, and check for any fuel leaks at the regulator and any other areas you may have disturbed. Reinstall the upper intake and all of its related hoses and wiring. Momentarily disconnect the battery to clear the computer's memory.

FUEL PRESSURE REGULATOR

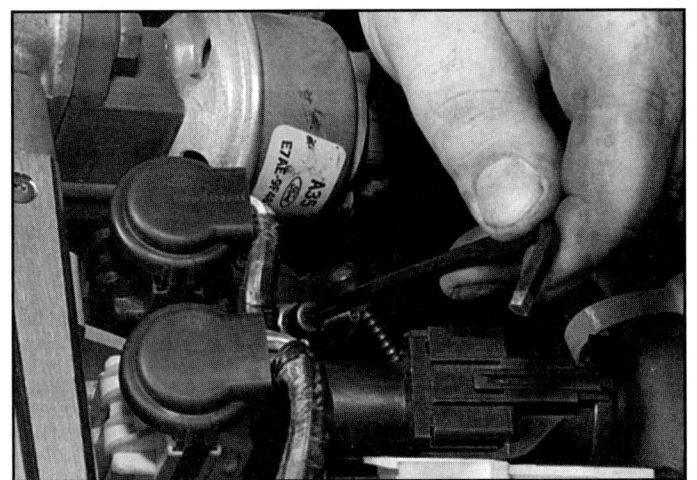

8. Install a fuel pressure gauge and start the engine. To adjust the fuel pressure, simply loosen the adjusting bolt's lock nut and adjust the fuel pressure as required. For best results set the fuel pressure to stock specs (38-40 PSI) and then experiment from there. Each engine package will require different amounts of fuel.

9. After the fuel pressure has been set, a deep 7/16 socket will fit over the adjusting bolt and can tighten the lock nut without disturbing the adjustment just made.

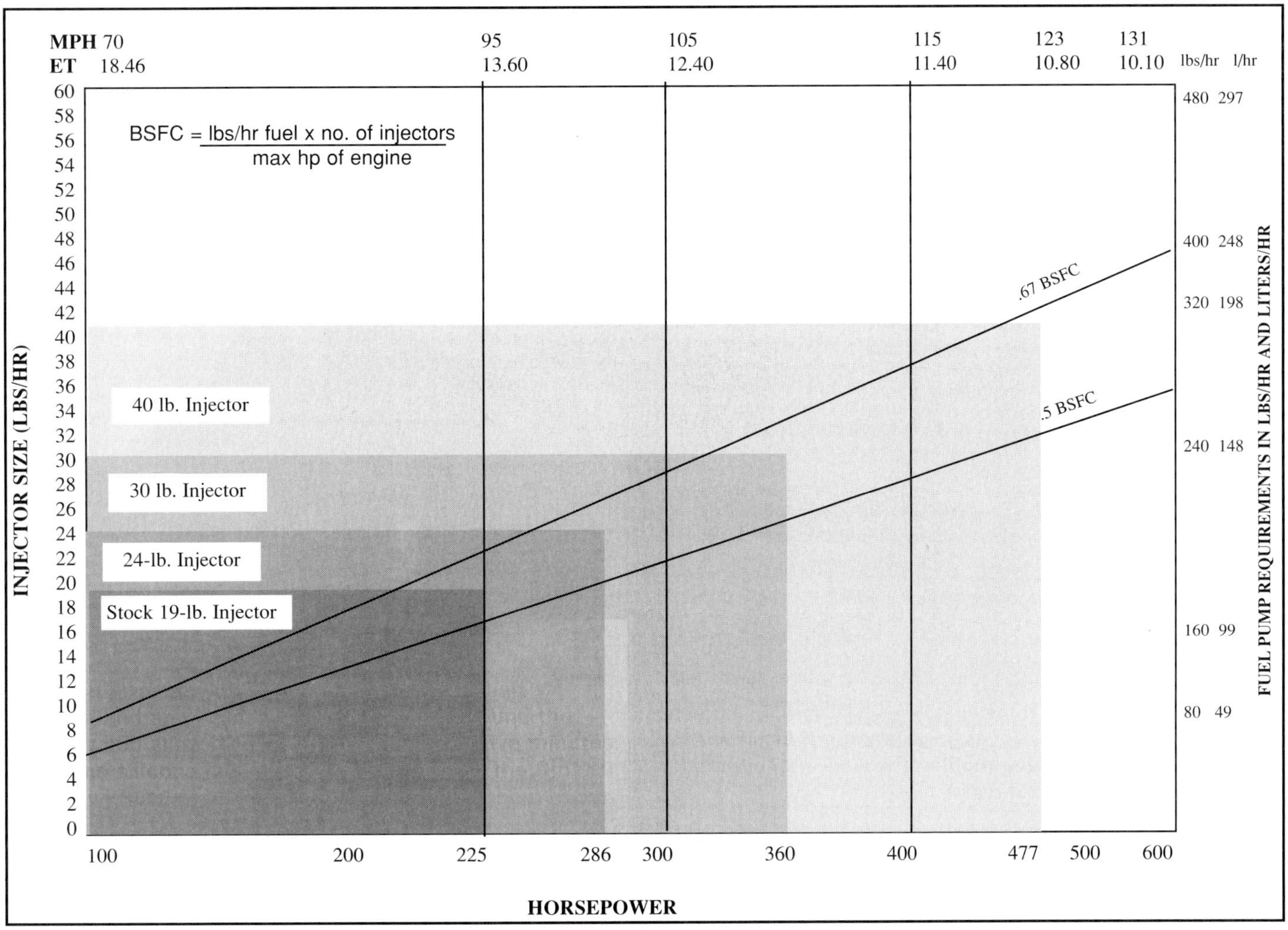

$$BSFC = \frac{\text{lbs/hr fuel} \times \text{no. of injectors}}{\text{max hp of engine}}$$

10. BSFC chart supplied by Kenne Bell.

75

HIGH VOLUME FUEL PUMP

One of the most common upgrades made to an EFI 5.0 Mustang is the replacement of the stock fuel pump with one of a higher flow rate. While the factory 88 LPH (liters per hour) pump is barely adequate for stock engines, the aftermarket 110, 155, and 190 LPH pumps can handle all but extreme high horsepower applications, which usually entail larger fuel lines and more than one pump, but we won't get into that discussion here.

The early EFI 5.0 Mustangs use a frame rail mounted fuel pump that is mounted externally from the fuel tank. These '84-'85 5.0s with the automatic transmission were the only Mustangs to use this type of fuel pump arrangement. Beginning in 1986, and the advent of the 5.0 SEFI engine in the Mustang, the fuel pump was relocated to within the fuel tank. There are several reasons why this change was made. For one, the addition of a true dual exhaust system would put the right muffler in the same place as the fuel pump and its bracket. The pump mounted in the tank can also benefit from direct access to the fuel, thus reducing the chance of fuel starvation in hard driving, the fuel pump would run quieter inside the fuel tank than bolted to the frame rail, and lastly, the pump can use the fuel itself as a cooling medium to prolong the life of the fuel pump. Access to the fuel pump requires removal of the fuel tank, but can be accomplished with your average hand tools.

Initially, the Ford Motorsport 110 LPH fuel pump was the only road to take, but the aftermarket has brought out 110 and 155 LPH pumps without the OE bracketry to reduce costs. These pumps require a few extra assembly steps during the replacement procedure, but function just the same once installed. The main cost savings comes from selling just the pump itself and not the related bracketry and hardware (money that could be spent elsewhere on your 5.0). With that said, we will tackle our fuel concerns with a new heavy duty 155 LPH fuel pump that can be obtained from any 5.0 performance dealer. This particular fuel pump is from Maximum Motorsports. We have included a list of common Ford replacement parts, and their part numbers, below, such as fuel filters and other incidentals you may wish to replace while installing the fuel pump.

The Maximum Motorsports HD 155 LPH fuel pump is just a replacement pump, not a complete drop-in replacement assembly like Ford Motorsport sells, but the few extra minutes of labor more than offset the cost.

FORD PART NUMBERS

pump to tank O-ring:	C0AF-9276-A
pump to tank lock ring:	C0AZ-9A307-B
fuel line clips, black:	N802239-S
fuel line clips, white:	N802241-S
fuel line clip, duck bill:	N802441-S
filler neck seal '81-'93:	E2DZ-9072-B
EFI fuel filter:	E7DZ-9155-A (FG-800-A)

HIGH VOLUME FUEL PUMP

1. It will be much easier to accomplish this task when the car is on "fumes." With a gallon or less in the tank (in the red zone of the fuel gauge), the tank will be much easier to handle. The tank when full will not only be heavy, but very dangerous. Support the rear of the car on jack stands and remove this 8mm bolt that retains the filler pipe support to the leading edge of the fuel tank. Remember, no smoking or open flames!

2. Place a small stool or milk crate (something with wheels is preferable) under the driver's side of the tank and remove the driver's side fuel tank strap retaining bolt. Unplug the fuel sender and pump harness (located just behind the bumper cover) at this time.

3. Remove the passenger side tank strap retaining bolt and let the driver's side of the tank rest on the stool or crate. With a helper holding the driver's side securely to the stool, grasp the passenger's side of the tank and rock it gently fore and aft, while pushing towards the drivers side. This will allow the tank inlet to slide off of the filler pipe. Center the tank on the stool to allow the removal of the fuel lines under the car.

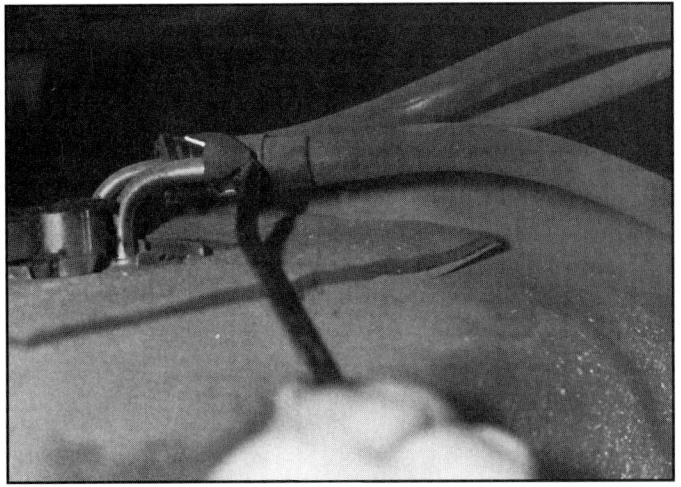

4. There are three lines that need to be removed on the Mustang, the pressure and return lines, and the vent line. Remove the vent line from the plastic vent first (not shown). The pressure line is retained by a white fuel line clip, similar to the ones used on the fuel filter. Using a blunt instrument, such as the cotter pin removal tool used here, pull the clip free of the fuel line.

HIGH VOLUME FUEL PUMP

5. Carefully twist and pull the line free. There may be some residual fuel still under pressure so beware of possible fuel getting into your eyes, etc.

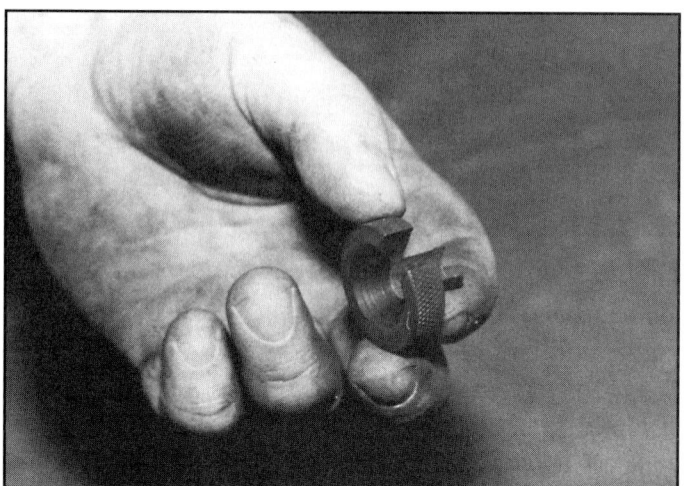

6. This is the specialty tool necessary to remove the return line, Ford tool number T82L-9500-AH. In the accompanying photos, we will show you how to use this tool.

7. As seen here, there are two tabs 180° apart from each other and 90° apart from the two external locating tabs of the "duck bill" clip (shown off of the car for photo clarity).

8. Insert the special tool so that the collar of the tool will spread these two tabs open and release the fitting from the fuel pump.

HIGH VOLUME FUEL PUMP

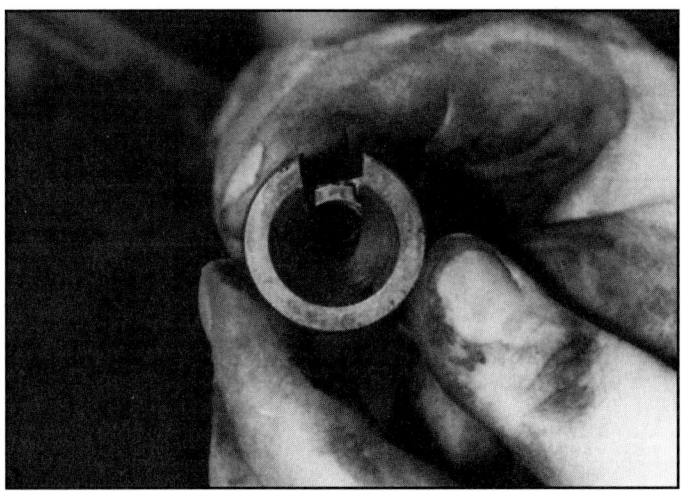

9. This is how it should look just before you remove the line from the fuel pump. Do not force the fitting, it is very possible you did not get both tabs released. It is also possible to use two small picks or screw drivers to release these tabs if you do not have the proper specialty tool for the job. Inspect the "duck bill" clip for any damage and replace if needed.

10. With all the lines removed, proceed to carefully roll the tank out from under the rear of the car and have someone help you place the tank on a comfortable work surface, such as the top of a garbage can or work bench.

11. Clean the entire area of any undercoating and dirt with high pressure air and carburetor cleaner before any removal steps are taken. This will prevent catching any loose undercoat on the new filter sock during installation and any possible loose contaminants from entering the fuel tank with the pump removed.

12. Disconnect the electrical wiring plug from the fuel pump and set it aside.

HIGH VOLUME FUEL PUMP

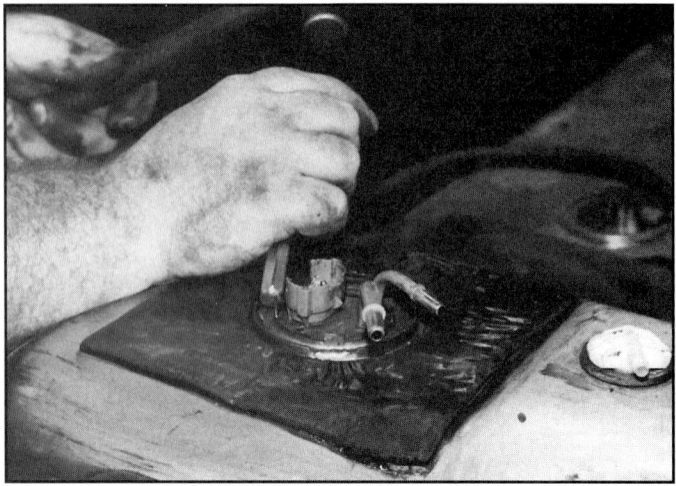

13. Use a brass drift punch (to prevent sparks) and a small hammer to unseat and loosen the retaining ring (counter-clockwise to loosen).

14. With the retaining ring removed, the pump can simply lift out with a twisting motion. There is an anti-fuel starvation cavity inside the tank to keep the fuel pump sock immersed in fuel at all times, and this device can make the fuel pump removal a tedious operation. Patience is a virtue here.

15. These fuel line clips and fuel pump O-ring are usually included with fuel pump kits, but we have included the Ford part numbers on page 76 for any future reference.

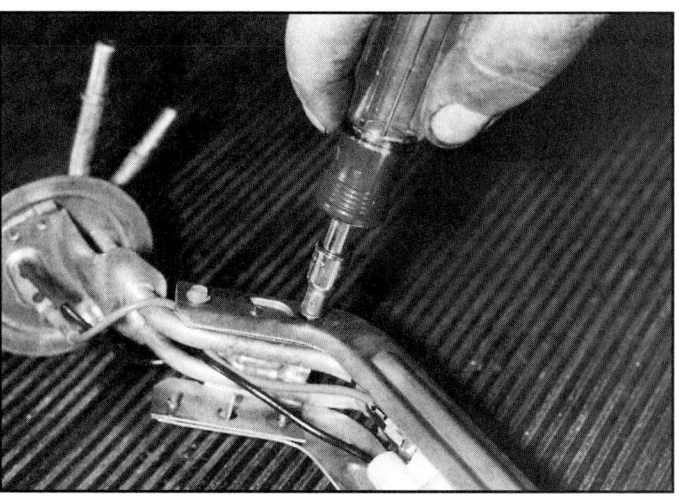

16. With the fuel pump on a clean work surface, remove the four bracket retaining screws from the fuel pump.

HIGH VOLUME FUEL PUMP

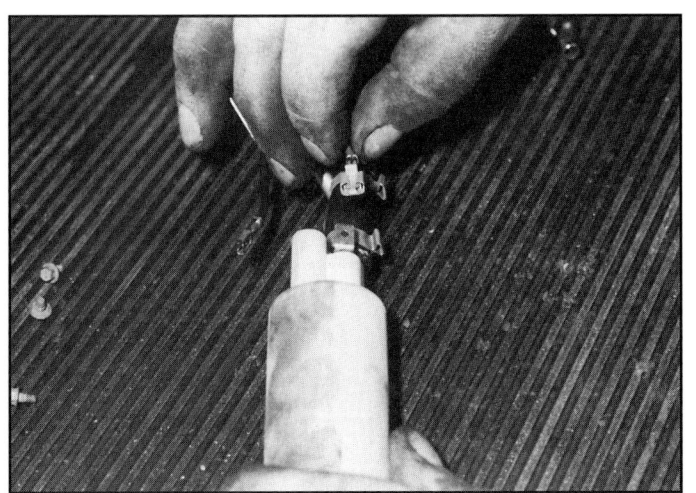

17. Proceed with removing the power and ground wires from the pump itself. The positive wire has a release tab that must be depressed before removal.

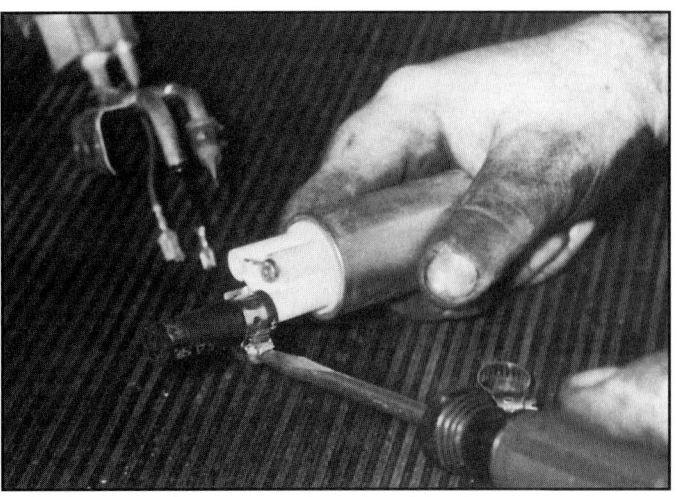

18. Remove the factory pump and insert the 155 LPH pump in its place. Our '88 5.0 had the factory fuel hose installed with crimp clamps; we replaced them with standard worm drive clamps. Inspect the fuel hose for cracks or deterioration.

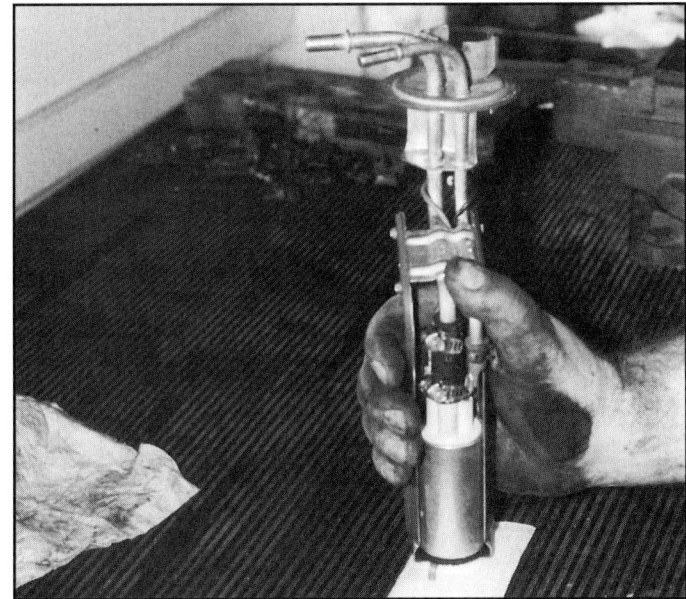

19. Reassemble the pump and bracketry and install the new fuel pump filter sock. The new sock is a high flow unit, but it must be indexed for proper location on the end of the pump by lining it up next to the old filter sock. Ensure the sock is fitted to the end of the pump securely by pressing down on the pump assembly until there is no gap between the filter sock and the pump body.

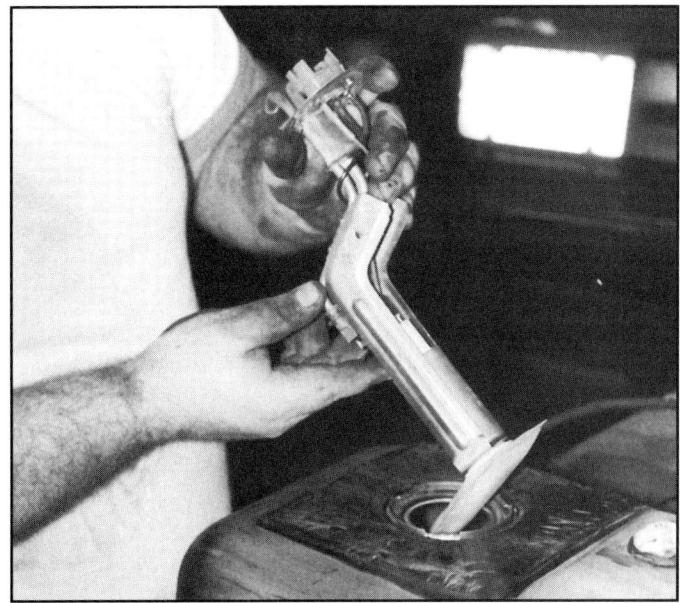

20. Install the new O-ring in the recess of the fuel tank and install the new fuel pump assembly and lock ring. Again, it will take some careful twisting and pushing to get the base of the pump/filter assembly into the anti-fuel starvation cavity in the tank. Reconnect the wiring plug at this time.

HIGH VOLUME FUEL PUMP

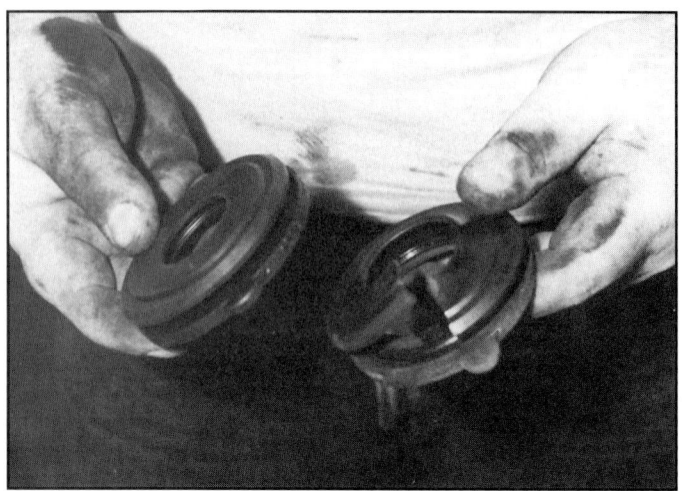

21. The fuel filler neck seal had torn (right), with years of fuel around it breaking down the rubber. A quick replacement insured a leak free fuel tank installation.

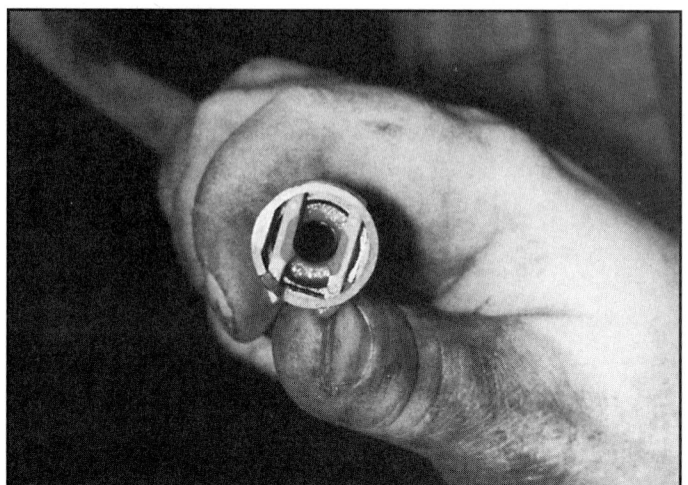

22. To correctly install the new fuel line retaining clip, ensure the two spring tabs are inserted into both the top and bottom holes, that the body of the clip is pointing away from the fitting, and the anti-removal tab is completely through the fitting to the other side.

23. With the tank back under the tail of our Mustang, we double checked all our fuel lines and test fired our '88 to check for any leaks. The fuel pressure was at a steady 32 lbs. with the engine running. Don't forget to drive straight to the gas station for a fill up.

NITROUS INSTALLATION

When the word "nitrous" is brought into a conversation, the first thing that people think of is "yeah, it makes lots of power until it grenades your engine." We aren't going to lie to you and say nitrous is easier on the engine than a mechanical bolt-on part, such as headers or roller rockers, but when installed correctly and used properly, nitrous will not hurt your engine. If you remember that the most important aspect of a nitrous system is an adequate fuel system for both the engine and the nitrous system, you won't have any problems. Let's discuss the basics of nitrous and how nitrous effects performance to help you gain knowledge about nitrous and how safe and useful it can actually be for your 5.0 Mustang.

Nitrous oxide was first discovered by an English scientist, Joseph Priestly, in 1772. Approximately a quarter century later another scientist from Britain, by the name of Humphrey Davy, mixed nitrous with oxygen, a mix of 66% nitrogen and 33% oxygen, and came up with relatively the same type of "laughing gas" that dentists use today. This term was coined due to the exhilarating effects that the gas had when inhaled.

Today's automotive nitrous is not like the nitrous used by dentists. Automotive nitrous has sulfur dioxide in it, which when attempted to be inhaled, gives off a "rotten egg" smell, hopefully deterring anyone from trying to inhale the gas. Nitrous oxide, when inhaled, does not break apart like when burned in an engine. The nitrous displaces the oxygen in your blood and withholds oxygen to the heart and brain, suffocating you in the process. Be very careful around nitrous when purging bottles or lines to not inhale it accidentally, and never inhale nitrous on purpose.

Though nitrous is under extreme pressure in the nitrous cylinder (usually from 600 to 1000 psi, which is roughly four to five times the amount of pressure in the typical air conditioning compressors "high" side), nitrous oxide is not a flammable gas, but an oxidizer. An oxidizer is a gas that promotes burning by adding its own oxygen, but is not flammable in itself. When the nitrous is entered into the engine's combustion chamber the oxygen and nitrogen separate, creating a lean condition due to the excessive oxygen. At this point fuel is injected to compensate for the excessive oxygen, which in turn makes more power by allowing an increase in cylinder pressure and volumetric efficiency, similar to blowers and turbos.

Nitrous has some advantages over blowers or turbos that should be addressed before you decide what type of power adder is really for you. First thing you might notice about nitrous is its relatively easy to swallow price over the comparable supercharger or single turbo kit. At roughly one fourth the cost of a typical supercharger this can sometimes be the main selling point of a nitrous system. Another factor to consider is the cooling effect of the nitrous itself. Nitrous has a boiling point of -129 degrees which cools the incoming air charge. For every 10 degree temperature drop you can expect to see a one percent increase in horsepower. So for a typical decrease of 60 degrees you will gain a six percent horsepower increase (13 horsepower on a stock 5.0) OVER the nitrous horsepower gains. Superchargers and turbos don't have this effect due to the fact they compress the air to fill the cylinders. If you remember your high school science class, compressing air makes the air hot, which loses efficiency. Nitrous is also an "on demand" power adder, meaning it only puts stress on the engine when you use it. Superchargers and turbos are running constantly whenever the engine is running. If you only use nitrous at the track for added power against another opponent, the 12 seconds of use during the pass at the track is much less detrimental than a continuous blower or turbo operation. Also, nitrous, when not in use, has no effect on fuel economy or emissions. About the only drawback to nitrous is having to fill the nitrous bottle when empty. Most speed shops can fill your bottle for

NITROUS INSTALLATION

about 25 dollars. Even if you used one bottle fill a week, it would take about a year and a half of bottle fills to equal the purchase price of a turbo or supercharger kit. This adjustable nitrous kit has jets for 50, 80, 100, 125, 150 and 175 hp levels. This allows you to start off with small amounts of nitrous and work your way up the ladder as you gain confidence and traction.

We obtained our nitrous system from The Nitrous Works. Our reasons for choosing TNW 5.0L Mustang adjustable kit were twofold. First, the plate style nitrous system is installed between the throttle body and the upper intake. This allows for better atomization after the throttle plate and does not require any drilling into the throttle body or intake charge hose, which could be seen by a technician when bringing the car in for warranty service. If need be, the nitrous plate can be removed for servicing very easily. Secondly, their patented safety system, which works in conjunction with their 5.0L Mustang kit, is an excellent deterrent for any form of nitrous mishap, including loss of oil or fuel pressure and sticking solenoids. The safety system should be considered a must when installing nitrous on an EFI car. Along with the basic adjustable system and the safety system, we also obtained a purge kit, IHRA bottle vent, and nitrous and fuel pressure gauges to accurately monitor and tune our system for best performance.

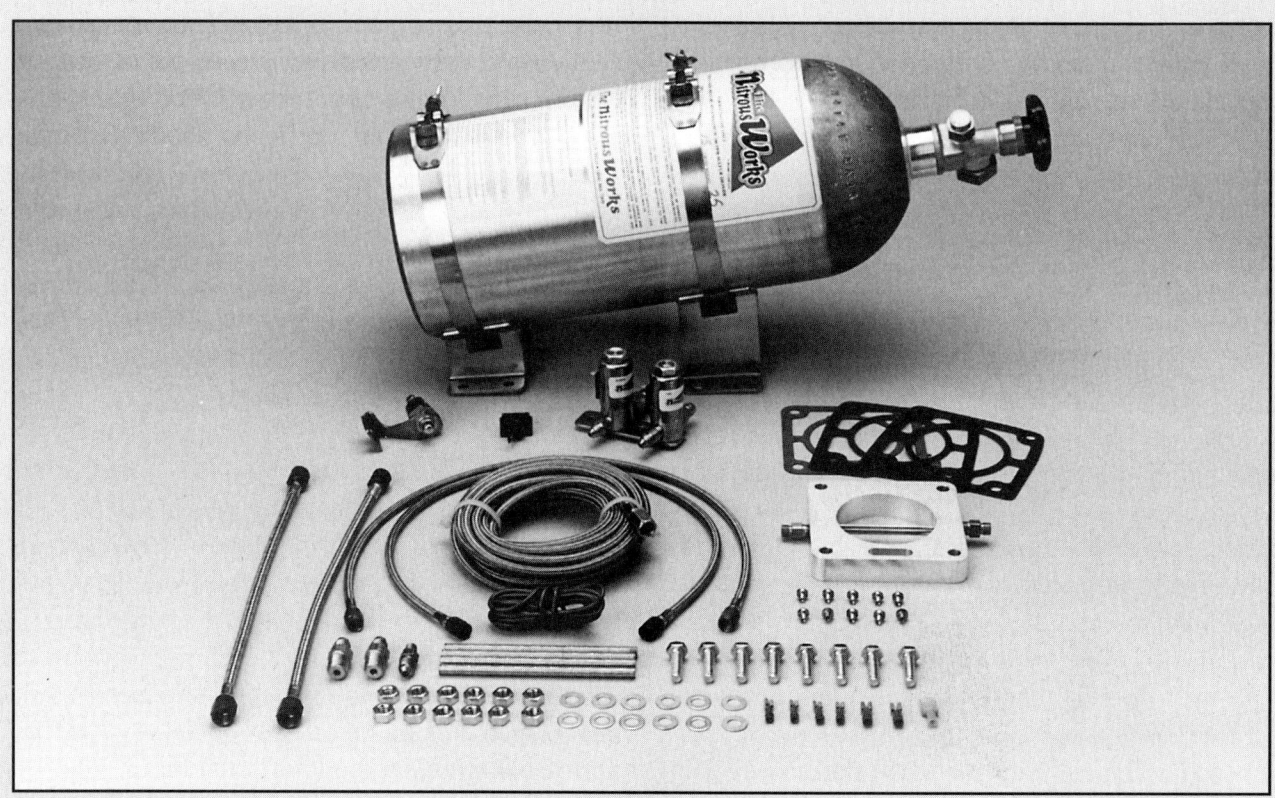

The Nitrous Works 5.0L EFI adjustable nitrous kit includes a filled 10 lb. bottle, bottle brackets, all necessary lines, nitrous and fuel solenoids, spray bar plate, hardware and instructions. Everything you need to go fast after an afternoon of wrench turning.

NITROUS INSTALLATION

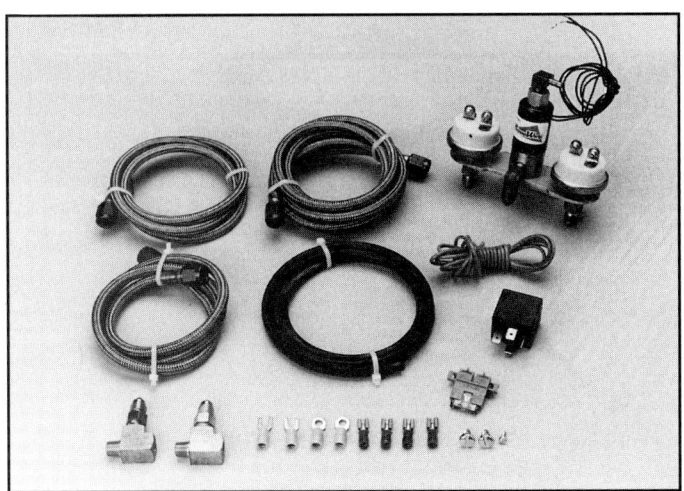

1. Along with the basic EFI system we added their patented safety system. This system includes two Hobbs pressure switches (one for fuel and one for oil), a nitrous vent solenoid, a relay to prevent melted switch contacts and all needed lines. This is the safest way to install nitrous on an EFI car. By monitoring your engine's fuel and oil pressures and monitoring the nitrous solenoid for sticking, this system will either shut the nitrous system down or vent the nitrous to the atmosphere instead of into your engine where severe damage could occur.

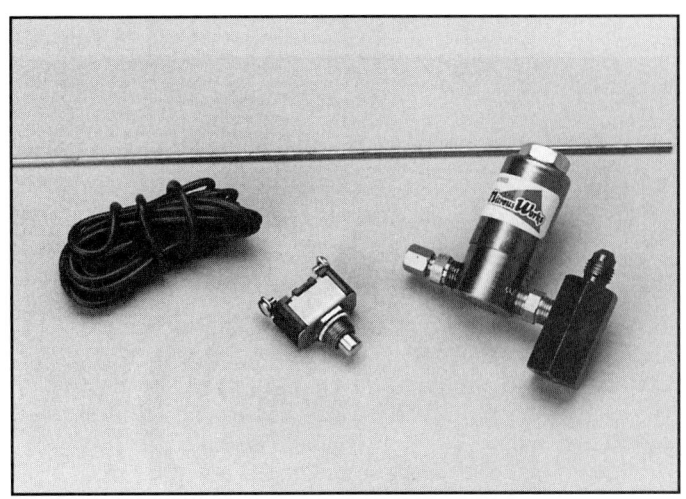

2. The Nitrous Works also supplied us with a purge solenoid to prevent rich conditions and power surges while using the nitrous system. This solenoid uses a momentary push button to purge all air out of the nitrous line (like bleeding the air from your brakes) before use for smooth, even power under nitrous usage.

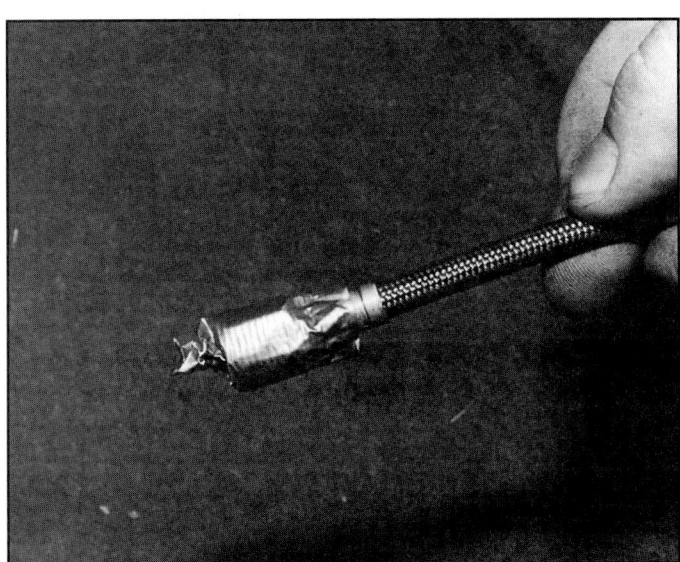

3. Before you begin running the main nitrous line from your trunk/hatch area to the front of the car, tape off the open end to prevent any foreign objects, dirt, etc. from entering the line during routing.

4. Ford was gracious enough to leave a body opening in the floor of the trunk/hatch area. Simply remove this plug, cut a hole in the center, and pass the nitrous feed line through this plug. Reinstall the plug to prevent water entry to the hatch or trunk.

NITROUS INSTALLATION

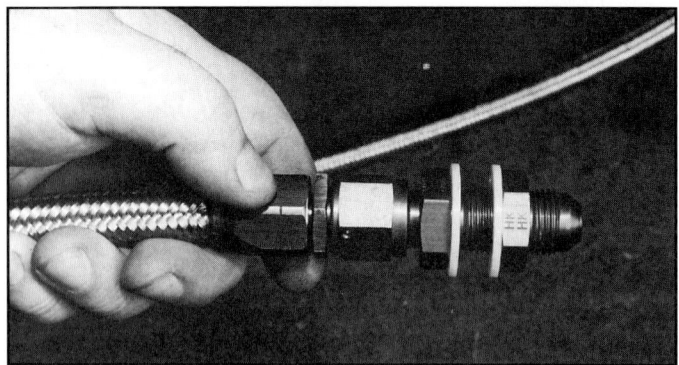

5. The Nitrous Works is an Aeroquip dealer, carrying their complete line of fittings and hoses. We used a -8 bulkhead fitting and a length of -8 braided line to vent the nitrous bottle to the outside of the car (IHRA legal), which is required by most tracks where the bottle occupies the same area as the driver (such as in a hatchback like ours). We used the trunk floor grommet directly across from the one we used for the nitrous feed line. Follow the factory routing of the fuel lines with the nitrous feed line, securing it every four to six inches with tie wraps. Make sure there is no slack or drooping line that can be caught on a road obstacle. Keep the line away from any suspension or exhaust parts. Route the line inside the passenger fender liner and into the engine compartment, following the route of the gas tank vent hose and into the engine compartment.

6. Remove the throttle body and EGR spacer and double nut the factory studs for removal. Once the two nuts are tight against each other, turn the inside one counter-clockwise to remove the studs. Do this for all four studs.

7. Install the new studs in the upper intake. The top ones will bottom and become tight, but the lower ones are threaded all the way through, requiring thread sealer. Install the lower ones even or slightly outward in relation to the top ones. Install the new gaskets and the nitrous spray bar plate. The plate is marked "top" and "N$_2$O". Ensure that the "N$_2$O" port is facing towards the front of the engine and that the "top" is facing up. The spray plate is a 70mm opening, perfect for our BBK 70mm throttle body and EGR spacer. Reinstall the throttle body and EGR spacer along with the nitrous full throttle switch. If you have custom race throttle bodies, TNW can make custom spray plates for your application. This spray plate configuration is the only way to legally run nitrous in the National Muscle Car Association's late model super modified class, as foggers are illegal.

8. Find a comfortable spot for your nitrous and fuel solenoids where the nitrous and fuel lines will reach the spray plate easily. The corner of the strut tower gave us good access to the spray plate, along with allowing us to hook up our lines without any interference. The purge solenoid is seen here already installed in series with the nitrous solenoid.

NITROUS INSTALLATION

9. Remove the tape from the end of the nitrous feed line and have a friend open the bottle valve for a few seconds to blow out any debris that may have been in the line from shipping or storage and install the line to the nitrous solenoid (or in our case the purge solenoid). TNW's tech department says if you must use a thread sealant use a liquid sealer with Teflon in it, such as Permatex sealer #14A, but do not use Teflon tape under any circumstances.

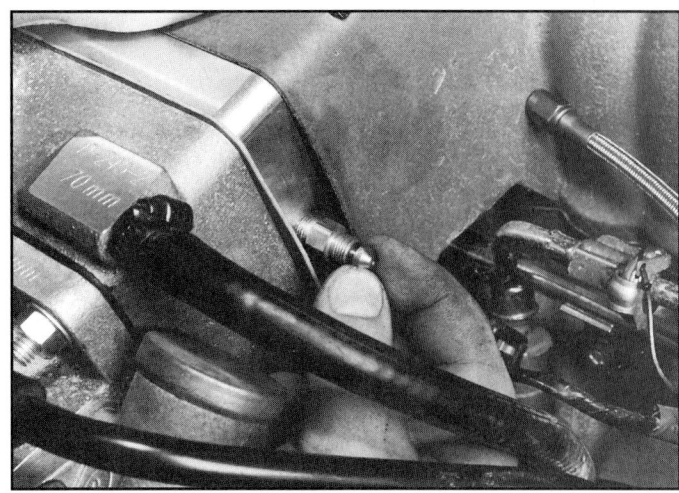

10. Insert the appropriate jet into the spray bar opening and install the line over the jet. We started off with the base 50 hp jets. We didn't want to be heroes and figured this would allow us to get used to running nitrous before we really poured it on to the Bowtie boys. The anodized line ends will help you determine which line you are working with. We used blue for nitrous and red for fuel.

11. Install the small lines from the nitrous and fuel sides of the spray bar plate to their solenoids on the strut tower.

12. Remove the factory Schrader valve from the fuel line (behind the alternator) and install the -4 fitting. This will allow the nitrous system to use the fuel from the fuel rail. Our Kenne Bell fuel regulator has a provision for this fuel "pick up" which will allow us to retain the Schrader valve for fuel system diagnosis. A nitrous system of this size requires additional fuel volume by means of a larger fuel pump. Adam Campbell of TNW suggests a 155 lph pump for a kit of this size.

NITROUS INSTALLATION

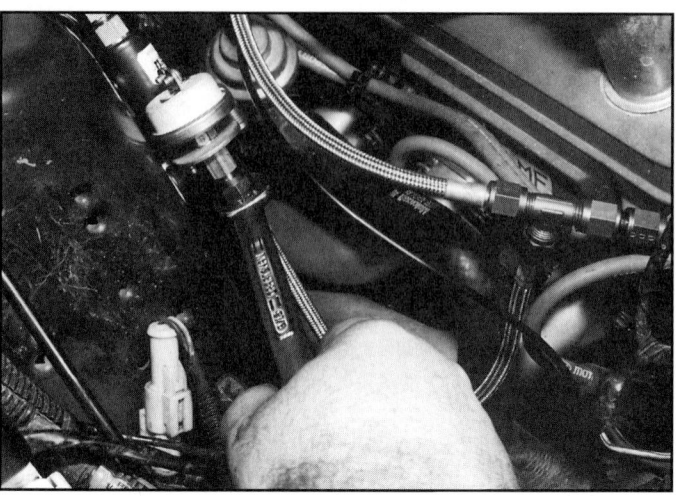

13. Run the -4 line from the fuel line to the adjustable regulator's "IN" port, then run another length of -4 line from the regulator to the inlet side of the fuel solenoid. Our optional fuel gauge is installed in one port of the regulator and the other is blocked off with an Allen plug. The regulator will allow fuel adjustments determined by the size of the nitrous and fuel jets used. The jets come with a jet card that shows you what pressures to set the regulator to. Do not set the regulator with the engine at idle, you will get a false reading. The car should be revved up to about 3,000 rpm and the throttle switch "blipped" (with the nitrous bottle off so only fuel flows) to check "flowing" fuel pressure. This will ensure that you have adequate pressure to the fuel solenoid under WOT conditions.

14. The safety system is installed on the face of the shock tower below the throttle body inlet hose. The -4 fuel line is connected to the fuel Hobbs switch and to the fuel line from the fuel rail by means of a double swivel and a "T" fitting.

15. The nitrous vent solenoid requires the installation of a "T" at the outlet side of the nitrous solenoid. One line runs to the spray bar plate and the other line (being connected here) goes to the nitrous vent solenoid on the safety system bracket. Install the flexible vent hose on the top of the vent solenoid and tie wrap it to vent away from the car safely.

16. The final safety system line to be hooked up is the long six foot -4 line to the oil pressure sending unit. Tie wrap it along the low pressure a/c hose along the back of the engine.

NITROUS INSTALLATION

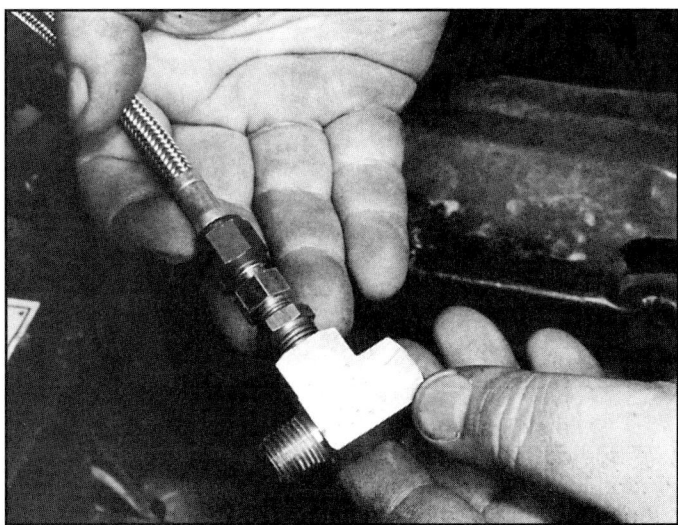

17. A 1/4" NPT "T" is used between the oil pressure sending unit and the engine to hook up the -4 oil pressure line.

18. The nitrous purge solenoid needs to have the purge tube routed to atmosphere. A common place is to route it up through a hood scoop or through the cowl vent, which is where we decided to run ours.

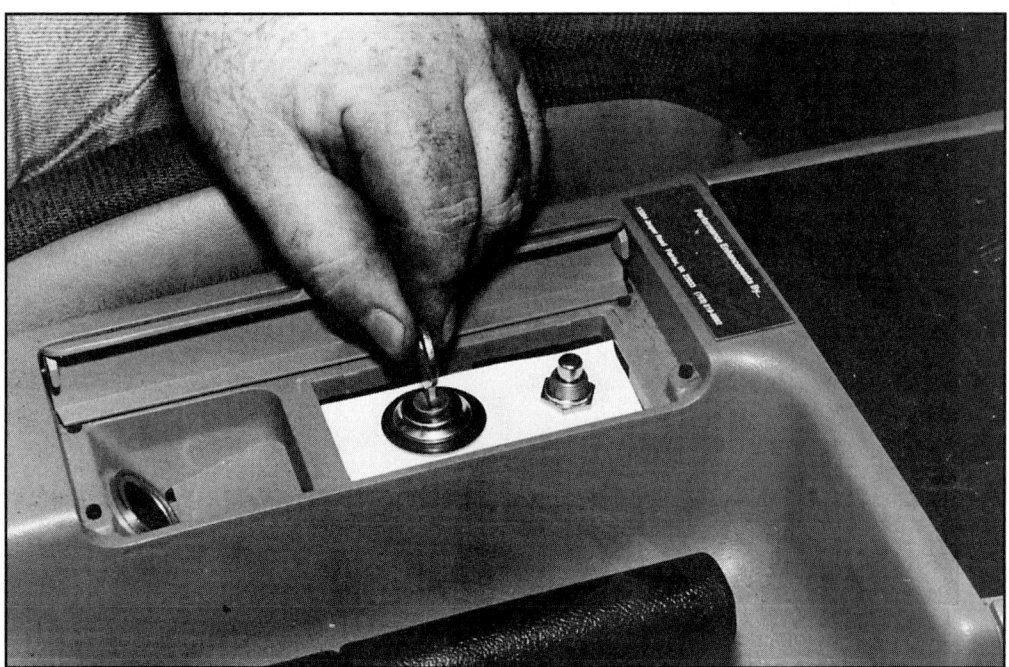

19. We decided to forgo the toggle switch for a safer and more controllable key switch (like the ones used in alarm systems). This will prevent anyone from wasting your nitrous or from hurting themselves or your car. The purge button can be seen next to the key switch.

SUPERCHARGER INSTALLATION

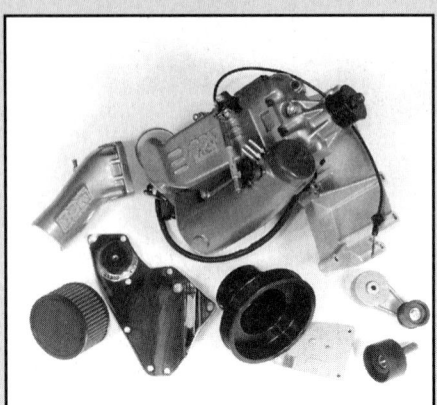

The hefty Instacharger comes assembled and mounted on its bracket, which makes it quite a load. The kit also includes all hard parts like the brackets, tensioner, crank pulley and inlet tube.

Superchargers are just about mandatory on 5.0 Mustangs it seems. You can't hit a cruise-in or drag strip without some 5.0 Mustang loping by with this mean sounding whine coming from under the hood. A supercharger or "blower" will definitely set you back some big green, but it will also set you back in your seat too! Today's blower kits, from manufacturers like Vortech, Paxton, Kenne Bell, Powerdyne, ATI, and BBK, all have been designed to be installed by the average tool toting home mechanic with a dose of common sense. What this means is that just about any of these manufacturer's systems could be installed on your Mustang in the course of a weekend, albeit a busy one.

For all of the applications listed above, all but the Kenne Bell and some ATI models use their own drive belt to drive the supercharger's impeller. The Kenne Bell and some of the ATI models use the OE serpentine belt to drive the impeller. The impeller is a vaned apparatus that pulls in air and compresses it within the blower housing. The compressed air is then delivered to the engine where auxiliary engine controls, usually supplied by the supercharger manufacturer, increase the fuel delivery, and if needed, reduce the spark timing to mix with the compressed air from the supercharger. In short, a supercharger mechanically increases volumetric efficiency by compressing the incoming air charge.

There are a few things you need to be aware of when choosing a supercharger. There are some units, like the Kenne Bell and BBK, that don't require an oil feed line from the engine, meaning they aren't lubricated by engine oil, but by oil within themselves. You need to decide before purchasing wether you want to use the engine's oil for lubrication of your supercharger or not. If you do, use a high grade oil filter, synthetic oil, and an external oil cooler. About the biggest fear I can think of is toasting your engine and sending all the crap in your engine oil through that feed line and taking out your three grand blower, or vice-versa. While it is a fear of mine it has been a reality to some owners, so think about that while you flip through *Super Ford* magazine and drool over the ads.

Most of the superchargers on the market today are also available with optional features to increase their horsepower level, looks (polishing or chroming), and warranty length. These are all options you should take into consideration when you are culling your facts for a future purchase. For example, if you can upgrade from a 6 psi supercharger to a 9 psi unit for just a few hundred dollars and still retain a warranty, it would behoove you to spend the additional monies now instead of purchasing the 6 psi unit and installing an aftermarket pulley to increase the boost, which would void your blower's warranty. Remember, let the buyer beware.

Installing a blower isn't like throwing a new fuel pump in your Mustang or bolting on that new shifter. A blower installation on average takes anywhere from 8-16 hours, depending upon the kit's contents and the type of blower being installed. So, if you're in the mood for a busy, non-stop weekend, turn on the answering machine, grab some handy snacks, a pitcher of iced tea, and get started bright and early Saturday morning.

The blower we decided to feature in this book is the relatively new BBK Instacharger unit, in a 9 lb. version. The steps are similar for most of the other manufacturers except for Kenne Bell, which mounts on top of the engine, and ATI, which mounts on the driver's side of the engine.

SUPERCHARGER INSTALLATION

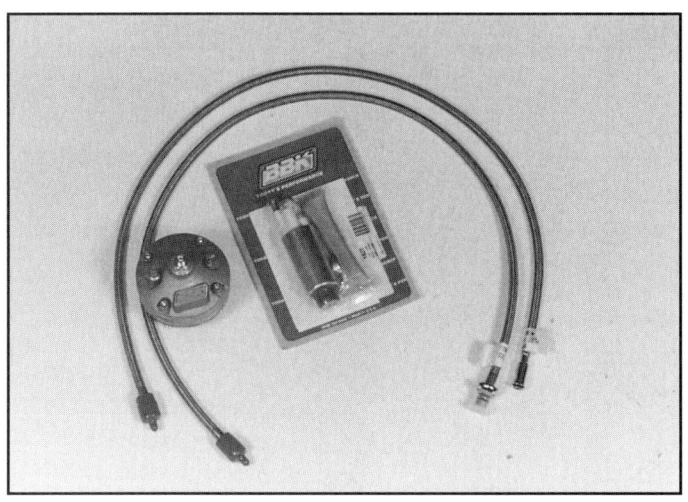

1. An adjustable FMU with braided lines and a 155 LPH in-tank pump are all part of the 9 lb. kit's fuel upgrades. Owners of modified 5.0s will likely need an in-line booster pump in addition to an upgraded in-tank pump, which BBK is developing. Buyers of 6 lb. kits don't get the pump.

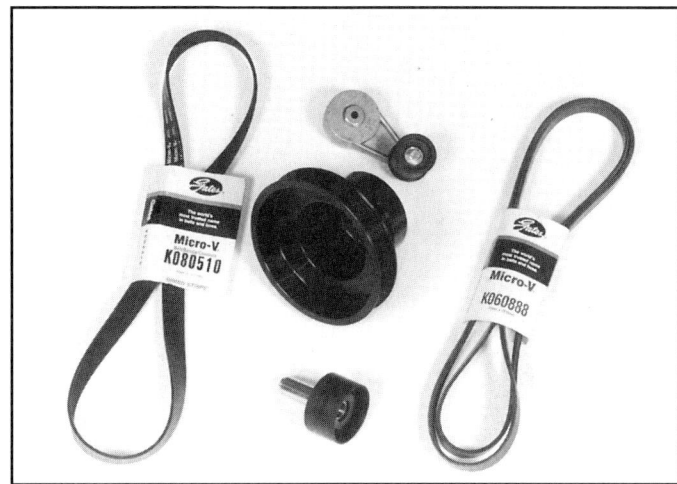

2. The blower drive gear is comprised of a one-piece billet aluminum crank pulley, underdriven on the accessory side and eight-rib on the blower side, an eight-rib blower belt, an idler, a spring loaded tensioner and a shorter serpentine accessory belt to accommodate the altered belt routing.

3. Begin by removing the air box, air meter and inlet tube as an assembly.

4. Follow up by removing the throttle body. The idle air bypass solenoid will be transferred to the BBK inlet setup.

SUPERCHARGER INSTALLATION

5. The Instacharger mounts in the familiar location; in place of the alternator. Since the BBK unit includes a new bracket, remove the alternator and Thermactor pump as a unit.

6. While there is plenty of working space, tap the adjustable FMU into the stock EFI return line using the AN fittings and braided lines included in the kit. Use Teflon sealant on the AN fittings to ensure a leak-free fit.

7. Here is the new crank pulley installed. As you can see it is pretty cozy with this car's aftermarket big radiator. The upper air conditioning hose seen here will snake under the installed supercharger, not over it.

SUPERCHARGER INSTALLATION

8. Next, install the Thermactor pump and alternator on the BBK bracket. We had previously fitted this 5.0 with Performance Parts Inc.'s Special Service alternator upgrade. The police alternator upgrade includes a special bracket, as its ears are a bit wider, which wouldn't let the alternator work with the BBK bracket. We used a 1-1/2" piece of tubing and a longer bolt (ARROW) to allow the police alternator to work with the bracketry. Also, it is important to connect the front portion of the Thermactor hose to the back of the pump before installing the blower.

9. Having the blower, Thermactor pump and alternator assembled on the bracket makes installation a definite two-person job. After lowering the blower down over the AC hose, have an assistant help by lifting the blower up while you install two of the bracket bolts into the cylinder head. For those running aluminum heads, don't forget the anti-seize. Follow up by installing the chrome blower idler/tensioner bracket.

10. While the BBK instructions suggest using ball-peen persuasion on the frame rail to gain clearance for the wiring harness, we chose to split the harness instead. We cut open the plastic tubing covering the harness and pulled out the wiring going to the heated exhaust gas oxygen sensor. We then ran the mass air and alternator wiring along the frame rail and the re-covered HEGO wiring behind the blower.

SUPERCHARGER INSTALLATION

11. Here is where the throttle body is relocated. The air filter and mass air sensor clamps go on in series in front of the throttle body. Idle bypass air and crankcase gas recirculation are achieved via three long rubber hoses, two traveling to the idle bypass solenoid and one to the valve cover.

12. Our two-core Be-Cool is as thick as most four-core radiators, so it made for tight clearance between the fan shroud and the Instacharger's eight-rib belt. We used a sanding disc to grind down the bottom and passenger sides of the fan shroud (ARROWS). The bottom requires the most clearancing, but with that done it bolted right up and cleared the belt and crank pulley. Ken Murphy at BBK says a three-core radiator will fit with no clearancing.

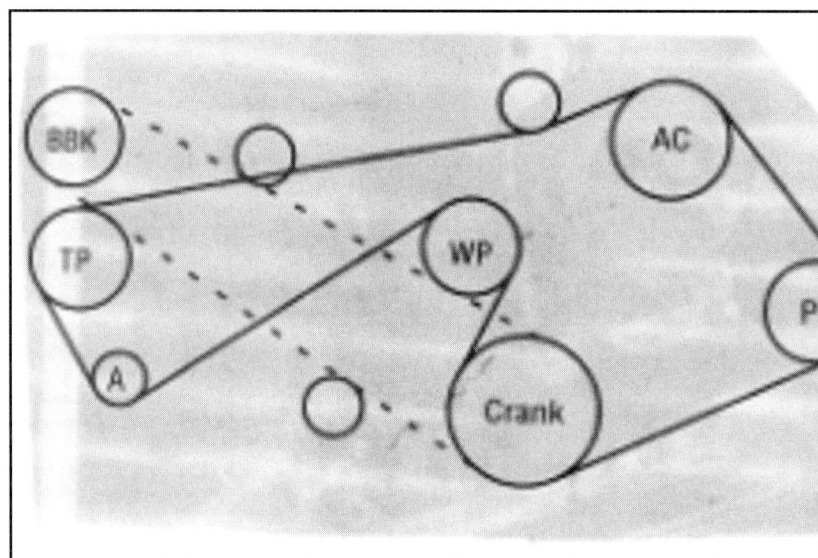

13. The accessory belt goes around the crank pulley, past the power steering pump pulley, over the AC compressor pulley, under the relocated tensioner, over the Thermactor pump pulley, down around the alternator pulley, over the water pump pulley and back down to the crank pulley. In front of the accessory belt, the blower belt routes around the crank pulley, under the idler pulley, over the blower pulley and on top of the tensioner.

SUPERCHARGER INSTALLATION

14. We have a length of heater hose clamped to the Thermactor pump and running straight out the back before we bolted on the blower. Then we reconfigured the check valves and hoses to this shape. The Thermactor Air Diverter transfers air to the heads or the catalytic converters depending on the situation. We moved its connection closer to the metal crossover tube and air injection tube. Then we formed an S-shaped link with the Thermactor Air Bypass, which diverts air to the TAD or to atmosphere, on its side in the middle. Then we clamped a metal tube betwixt the heater hose coming from the pump and the hose going to the TAB. All this manages to clear the header and the bypass valve at the rear of the blower.

15. Once again we deviated from the instructions on installation of the fuel management unit. BBK's instructions suggest installing the FMU inside the fender well. We, however, wanted direct access to the adjustable unit, so we placed it on the shock tower before proceeding with the inlet tube installation. We made sure to tie-wrap the braided lines away from the header.

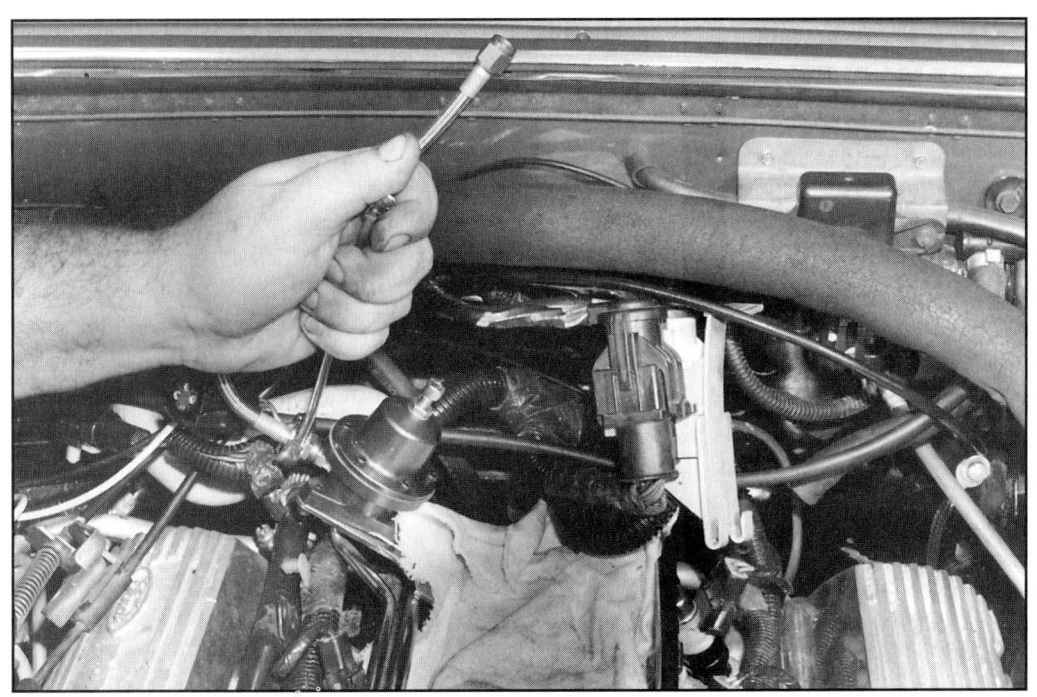

16. Prior to the blower installation we installed one of Kenne Bell's high-quality adjustable fuel pressure regulators (shown in the fuel section of this book). These billet aluminum, rebuildable units will withstand up to 100 lbs. of rail pressure, which makes us feel a lot better than forcing FMU-type pressure past the feeble stock regulator. As it turned out, the Kenne Bell unit served up more than durability and off-boost tunabilty. We used its auxiliary port and a length of braided line to relocate our fuel pressure sending unit, as Schrader valve access is greatly reduced by the Instacharger.

SUPERCHARGER INSTALLATION

17. We thought we were on the home stretch until the throttle linkage on the BBK elbow conflicted with our taller Motorsport polished valve covers. The simplest fix was to install one of BBK's 3/8" phenolic spacers, which gained us the necessary clearance.

18. After jumping the linkage hurdle, we installed the idle bypass valve. Again, our taller valve covers were in the way – crimping the hoses to and from the bypass. We may relocate them in the future, but for an application running stock valve covers, such clearances are not an issue.

19. Here is the finished product. Note the upper radiator hose is quite chummy with the blower pulley. It is necessary to trim about an inch from the hose to get it to clear the pulley. We ended up trimming an inch and a half because of our big radiator, but the hose has not rubbed once. We would also recommend installing platinum spark plugs before installing the blower, as the first and second plugs on the passenger's side will have to be changed from underneath with the blower in place. While initially not as familiar and sexy as the common centrifugal look, the Instacharger look quickly grew on us, and we did get a "Man that's a mean lookin' blower" comment, so different might just be better on the burger stand scene.

V. TRANSMISSION TECH

**CLUTCH REPLACEMENT
INSTALLING A PERFORMANCE SHIFTER
SHIFTER BUSHINGS
TRANSMISSION OIL COOLER**

CLUTCH REPLACEMENT

Sooner or later we all get it; no, not an IRS audit, but a slipping or chattering clutch. Whether you drag race your car or not it doesn't matter, of course drag racing will accelerate the amount of clutch wear, but when the clutch comes to the end of its service life it spells grief for the driver. Several different maladies can render a clutch ineffective or useless. Oil penetration of the clutch surface is one concern. If you have a rear intake oil leak or a rear main seal oil leak both can soak your clutch with engine oil to the point of replacement in no time at all. Simple mileage wear will also cause problems, as well as incorrect clutch adjustment.

One-piece adjusters are great for racing environments, but if they aren't adjusted right they will cause the clutch to go "over center" and push the pressure plate fingers too far, damaging the clutch and pressure plate. If this isn't enough to scare you then take a look at the front bearing retainer that comes on the factory T-5. It is made of aluminum throughout and this is where the clutch release bearing rides, eventually scoring up the sleeve of the retainer and jamming the bearing.

We will address these complaints in our clutch buildup on this '88 5.0, which barely made it to the shop with the clutch problems that the car exhibited, by replacing our intake gaskets, rear main seal, front bearing retainer, clutch disc and pressure plate, and readjusting our clutch cable. A call to Centerforce, makers of the infamous Dual-Friction clutch assembly, with a quick description of our concern and we heard "another retainer job" on the other end of the phone, meaning they have seen this thousands of times before and suggested we replace the shift fork, pivot ball, front bearing retainer, and install a new clutch. Follow along as we slide under our 5.0 for some serious clutch maintenance.

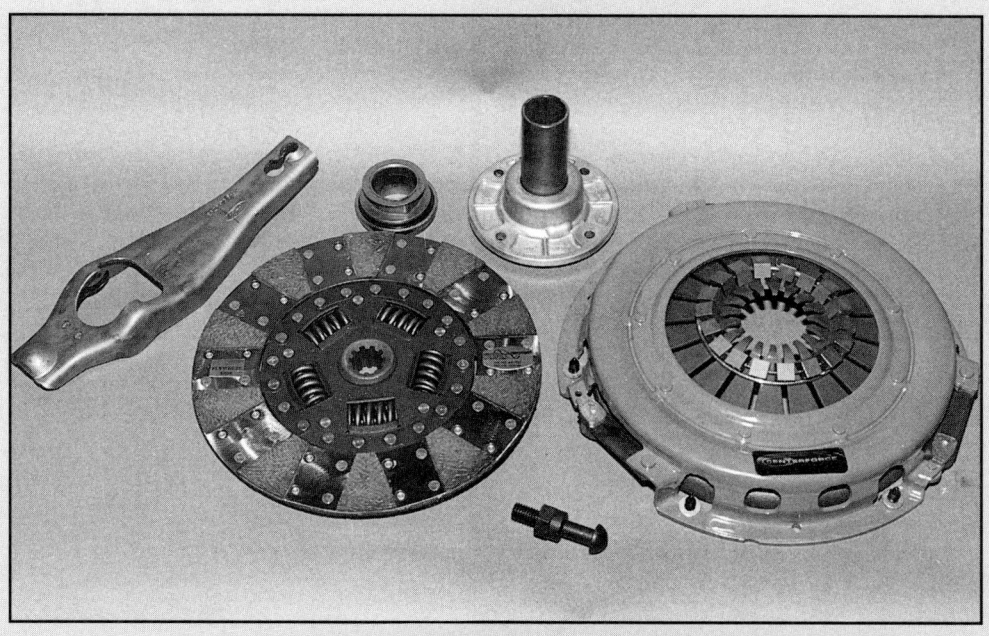

Our selection of clutch components include a new Centerforce Dual Friction clutch and pressure plate, Centerforce release bearing, a D&D Performance steel bearing retainer, and factory Ford clutch fork and pivot ball.

CLUTCH REPLACEMENT

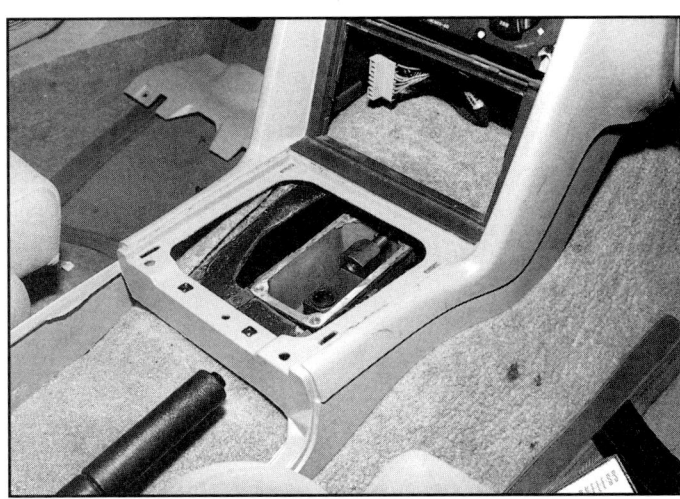

1. You will need to remove the shift knob and boot to access the shifter assembly. Once the boot and knob are free, remove the shifter handle from the shifter turret, or remove the shifter assembly altogether so the shifter can be inspected and/or rebuilt.

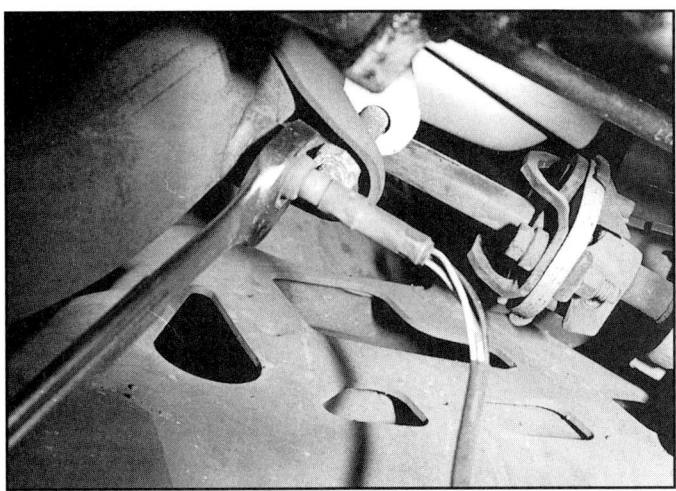

2. Before you can loosen the H-pipe to header nuts you will have to remove the two HEGO sensors from the H-pipe, just below the header flange. Disconnect the electrical connector and use a 7/8 inch wrench, HEGO socket, or "weather head" socket to remove the HEGO sensor, then remove the four header flange nuts.

3. Where the H-pipe and muffler tubes meet are two flanges that will have to be unbolted. Remove the four nuts and push the mufflers rearward to free the balljoint fitting. Let the mufflers hang free for room to remove the H-pipe.

4. The last thing you will have to accomplish to remove the H-pipe will be to remove the downstream thermactor air tube from the H-pipe. Loosen the clamp around the joint and slide the clamp out of the way. A torch should be used to heat the air tube until you can slide the tube off. If you don't have access to a torch you can slit the side of the joint to ease the tube off or you can simply hacksaw the tube off and reinstall the tube with a short length of high temperature hose later. We have also removed our starter at this time (arrow).

CLUTCH REPLACEMENT

5. Once the air tube is free of the H-pipe, place a pry bar between the H-pipe crossover tube and a solid object on the body, such as the cross member, or transmission, and pry the H-pipe back until it is free of the headers and free of the rubber retainers on the cross member and remove the H-pipe from under the car. Mark the driveshaft and companion flange on the rear end with touch up paint or a grease pencil and remove the driveshaft. A 12mm 12-point socket will be needed for the drive shaft bolts. Heating the companion flange with a torch will ease the removal of the driveshaft bolts, which are retained with thread locking compound. A tailshaft cap will be needed to prevent fluid from exiting the transmission.

6. Remove the two crossmember to transmission mount nuts. Place a floor jack under the transmission and jack up the transmission. Remove the crossmember bolts and the crossmember. Remove the four transmission to bell housing mounting bolts (arrow) and disconnect all electrical wires and the speedometer cable.

7. With one person in the car and another under the car, run a length of rope around the tailshaft of the transmission and have the person inside the car pull up on the transmission while you slide the transmission out of the bell housing and to the ground. Be careful, the transmission is quite heavy. This is one of the safest ways to remove it without a transmission jack.

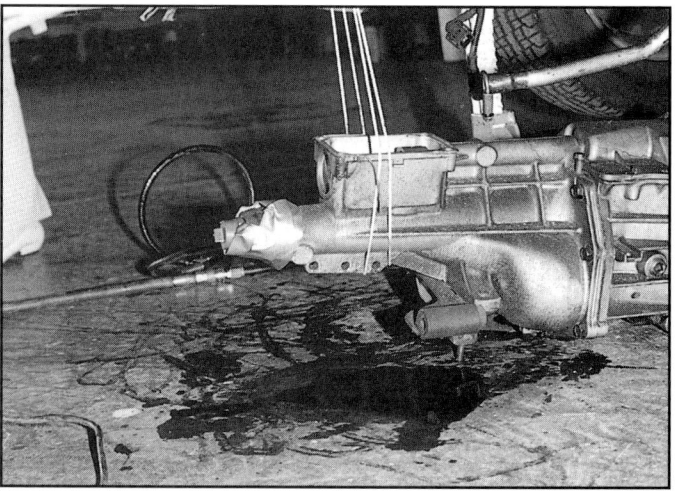

8. Once the transmission is on the ground remove the rope and transfer the transmission to a workbench, or in our case to an area to be degreased.

CLUTCH REPLACEMENT

9. Disconnect the clutch cable and remove the bellhousing. Six bolts attach the pressure plate to the flywheel. As you loosen the pressure plate the clutch disc will start to slide out from between the pressure plate and flywheel.

10. The clutch fingers show evidence of the release bearing sticking and/or the clutch going over center (arrow).

11. An impact gun or a friend holding the crankshaft with a breaker bar will be needed to remove the flywheel mounting bolts. Remove the engine to transmission dust shield at this time too. A trip to the machine shop will be needed to surface the flywheel.

12. Though we haven't degreased the area yet our new rear main seal is easily installed by gently tapping it in place with the tip of a block of wood. It would be wise to add some non-hardening sealer around the seal's edges before installation. You can see the evidence of the intake leaking from the top of the block.

CLUTCH REPLACEMENT

13. The inside of the bell housing was saturated with oil, which certainly led to the demise of our clutch components.

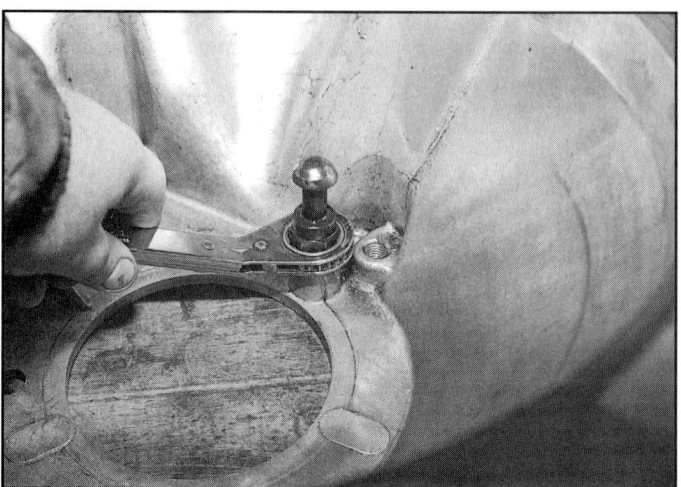

14. After a complete cleaning our bell housing is ready for another 70,000 miles of service, except for our clutch fork pivot, which was promptly replaced by a stock Ford piece.

15. Our rear main and intake leaks had taken their toll not only on the clutch but on the outside of our transmission too. A thorough cleaning with engine degreaser made our transmission look like new and easier to work on. Everyone likes to work on clean parts, don't they?

CLUTCH REPLACEMENT

16. Four bolts retain the front bearing retainer assembly to the transmission case.

17. Temporarily reinstall the lower bolt to prevent the transmission fluid from leaking out. A short block of wood to prop up the front of the transmission case won't hurt either.

18. The new D&D steel retainer is shown at left. Transfer the original bearing race and selective shim (found beneath the bearing race) to the new D&D piece. The oil seal is already installed. Return the new retainer to the transmission case with a light layer of silicone as a sealer between the two parts. Thread sealer should also be used on the retainer bolts as they protrude into the case.

CLUTCH REPLACEMENT

19. Our stock flywheel was resurfaced to ensure optimum clamping force with our new Centerforce Dual Friction clutch assembly.

20. Install the flywheel back onto the crankshaft, using thread sealer on the flywheel bolts (don't forget the dust shield first). The flywheel only bolts on one way so ensure all bolt holes are lined up before inserting the bolts. Place the clutch disc and centering tool into the end of the crankshaft and bolt down the pressure plate over the clutch disc. Once the pressure plate is tight you can remove the centering tool.

21. Our cleaned bell housing now sports a new pivot stud and clutch fork along with our new release bearing. Reinstall the bell housing to the engine block and reinstall the starter. The remainder of the operation is simply the reverse of removal. When everything is reinstalled either adjust your one piece quadrant as per the manufacturers instructions, or if you have a stock clutch quadrant simply follow the instructions in your shop manual or owner's manual for proper clutch release action.

INSTALLING A PERFORMANCE SHIFTER

Not too many people think about the shifter in their 5.0 Mustang, and if they do, it's probably right after the shifter handle comes off in their hand during a speed shift. Shifters really don't get any respect when you think about it. They are beaten relentlessly, getting shoved back and forth trying to shave that last tenth of ET off your Mustang. For a long time there were no options to upgrading the shifter on your T-5 five speed transmission. Ford commissioned Hurst, famous for their shifters of the Musclecar era, to design a performance shifter for the now-defunct Mustang SVO. This Hurst T-5 shifter found its way into the Ford Motorsport catalog and into thousands of Mustangs. The Hurst shifter, though, used a thin aluminum base like the stock shifter's stamped-steel base and was prone to the same flexibility problems of the stock shifter.

Since that first Hurst shifter for the 5.0 Mustang, several companies have tried to improve on the Hurst design with different handles, stronger stop bolts, and solid mount bushings to replace the rubber isolator between the handle and the shifter base. All of these items worked fine to a certain degree, but a true performance shifter that could take a pounding at the strip from round to round was needed, and that void was filled by Pro 5.0's Power Tower shifter.

Pro 5.0 has redrawn the lines of battle when it comes to a replacement shifter assembly for the sometimes fragile T-5 manual transmission. Their all new shifter design not only incorporates the "must have" shifter stop bolts to prevent shift fork damage, but the body of the shifter, as a whole, is designed around an abusive lifestyle of speed shifts. The tower design of the shifter assembly allows the force of a severe shift to be transmitted to the tower, thus spreading the force around the shifter and not just in one direction. The tower is made from surgical steel, like those found in most knee surgery replacements, while the stop bolts are made from aerospace strength materials. The shifter also deletes the rubber isolator that the stock piece uses for firmer, more positive shifts. The shifter is available in several different handle lengths; and custom handle lengths or angles can be made upon request. Our donor car has the original T-5 in it and has never been speed shifted (the owner simply can't afford the time and money involved in having his car down for transmission service). Once we complete our installation we are sure he will feel much more confident the next time he grabs that shifter.

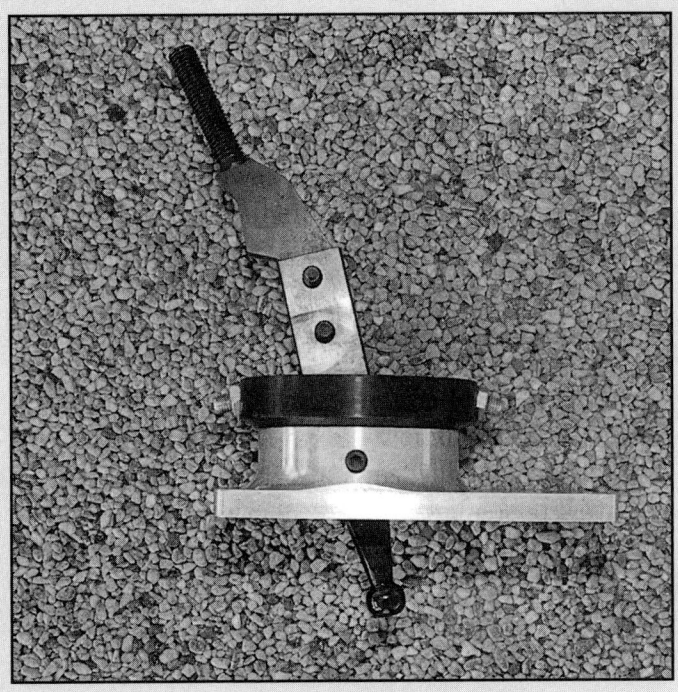

The Pro 5.0 "Power Tower" shifter will let your T-5 grow old gracefully. No need to put a zipper on your transmission for easy service anymore.

INSTALLING A PERFORMANCE SHIFTER

1. Our donor car had a strange but effective layout of a short stick with a T-handle on the end of it, somewhat awkward for us at first but the owner claims to love it.

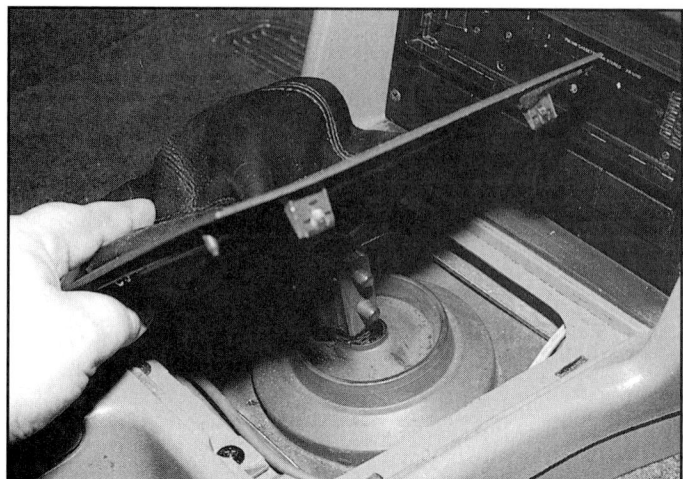

2. Unscrew your shift knob assembly and remove the shift boot bezel assembly by prying up at the four corners to release the retaining clips.

3. Our short shifter stick was an aftermarket piece installed a few years ago. While these units do give more precise shifting by eliminating the rubber isolator, the shifter stop bolts are a necessity for the T-5. Remove the two attaching bolts to remove the shifter stick.

4. The four small 5/16" bolts that hold the weather seal in place are removed next. The two in front can be removed with an ignition wrench to prevent disassembling the complete console, or a small "tab" can be cut away from the console to access the bolts.

INSTALLING A PERFORMANCE SHIFTER

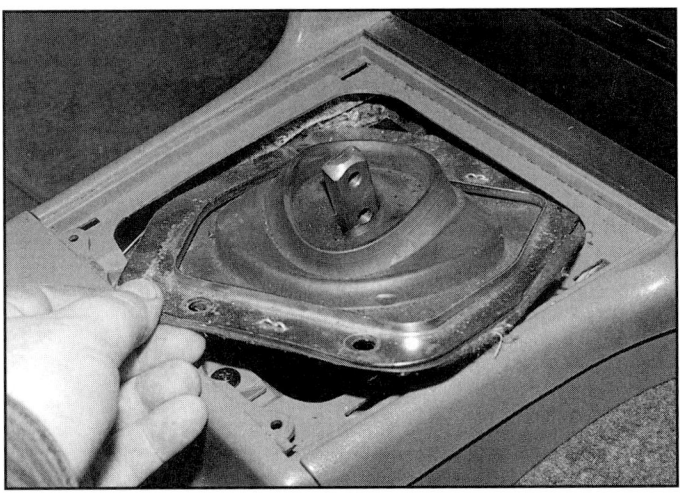

5. Once the four attaching bolts are removed, the weather seal can be carefully worked out of the console opening. Start with one corner of the seal and work it around to remove it.

6. With the stock shifter exposed, remove the four 13mm bolts attaching the shifter base to the tailshaft of the transmission and carefully pry the shifter base free from the transmission.

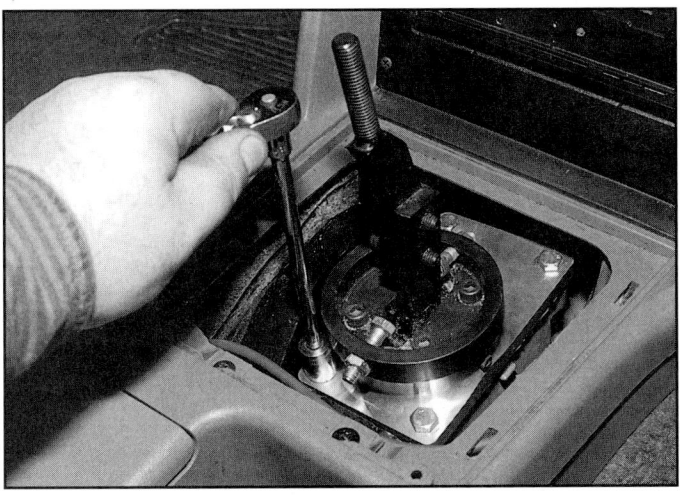

7. Clean away any old silicone from the tailshaft's opening and apply a new bead of silicone around the opening and allow it to "skin" over in about five minutes. Once the silicone has started to skin, install the Pro 5.0 shifter with its new longer bolts and tighten them evenly.

8. Once you reinstall your weather seal, shifter boot and handle no one will ever know what you have waiting and ready for that grudge night at the local track. We can nail third gear now with all our might and not wince every time we throw the shifter forward.

SHIFTER BUSHINGS

Upgrades, especially low-buck upgrades, are always a welcome sight in the arena of high performance cars like the 5.0 Mustang. Sure, we wish we all could spend our hard earned dead presidents on a supercharger or some 18" aluminum wheels, but in reality we have to worry about the light bill, food for the table, and other necessities to keep our family secure and content (and from having your significant other toss you out on your ear). But, we're all car guys and gals and we couldn't be comfortable without giving up some small chunk of our meager paycheck to our beloved 5.0 Mustang. So what do you do? You could save and save for months on end until you have enough to buy what ever precious bauble you are drooling at in the performance catalog du jour. Or, you could lay down that small chunk of change for something you do need but won't take a year to pay off. Items such as under drive pulleys, stripe kits, short throw shifters, auxiliary gauges, alarm systems, window tinting, billet pedal covers, head and tail light covers or treatments, throttle bodies, shorty headers, K&N Filters, etc. Every one of these items can be bought or installed for under $200, and some for under $100! And every one of these items I mentioned will either improve the looks, performance, security, and/or resale value of your Mustang.

Steeda Autosports, a name synonymous with 5.0 speed parts across the US, has inexpensive yet successful performance upgrades for the 5.0 Mustang. These include the Steeda stainless steel caliper guide sleeves, Urethane bushing kits, and their even more popular stainless steel shifter bushings. These bushings take the place of the factory rubber isolator setup and improve shifting control tremendously by making the connection between the shift handle and the shift plate rigid, offering a precise shift feel for accurate shifts every time. Not only do these bushings fit the factory T-5, but they work perfectly on the Tremec too. Though Stainless Steel shifter bushings have been on the market for some time, we felt it would benefit first time 5.0 owners on a tight budget to see how easy and inexpensive items like these bushing are to install.

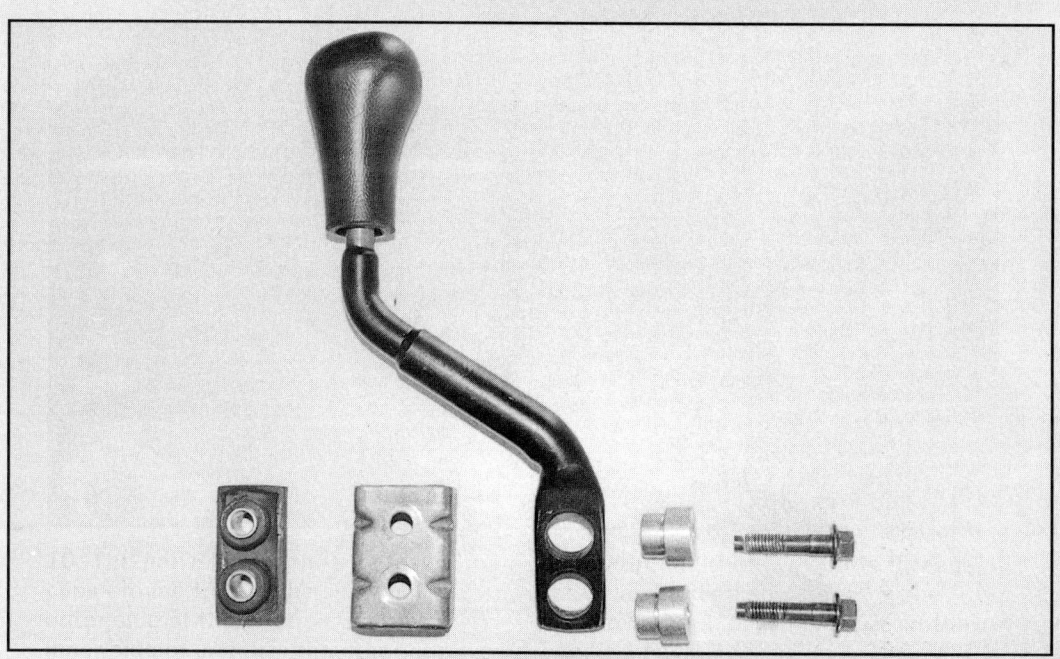

To the left of the shift handle is the factory rubber bushing and to the right of the shift handle is the replacement Steeda stainless steel shifter bushings. The bolts to the right of the Steeda bushings are the stock bolts.

SHIFTER BUSHINGS

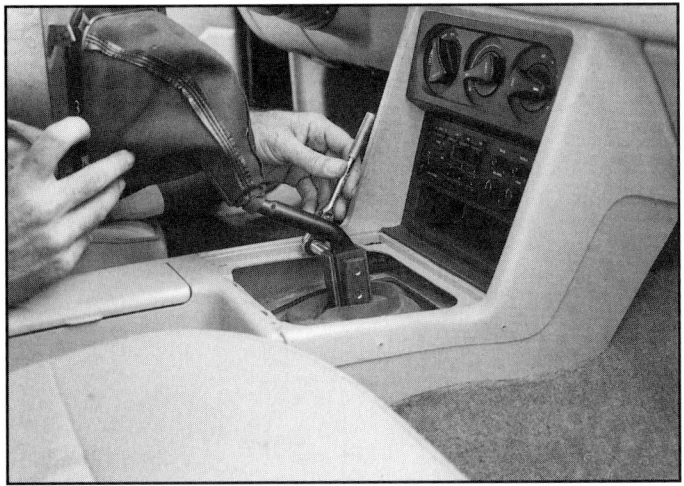

1. Gently pry the shift plate up at the corners and pull the boot up the shaft of the shift handle to gain access to the shifter bushing. Remove the two bolts holding the bushings and the shift handle.

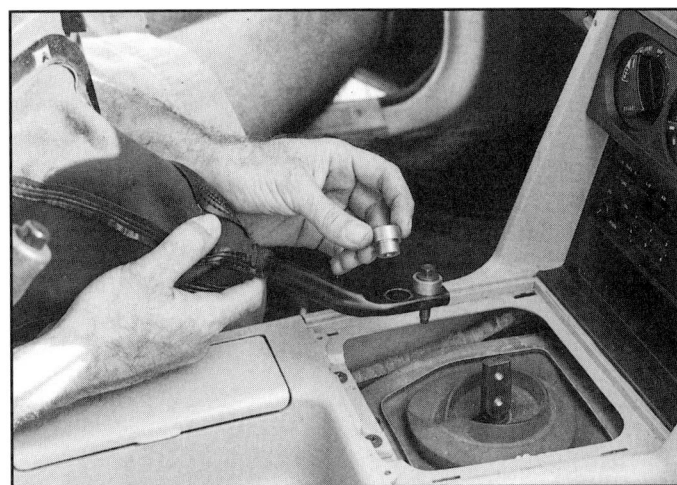

2. Install the new Steeda shifter bushings into the shift handle, after removing the stock bushing from the handle, then place the attaching bolts in position through the stainless steel bushings.

3. Tighten the bolts evenly to 23-32 lb.-ft. When reinstalling the bolts with the Steeda bushings, ensure that Lock-Tite is used to prevent the possibility of the bolts loosening.

SHIFTER BUSHINGS

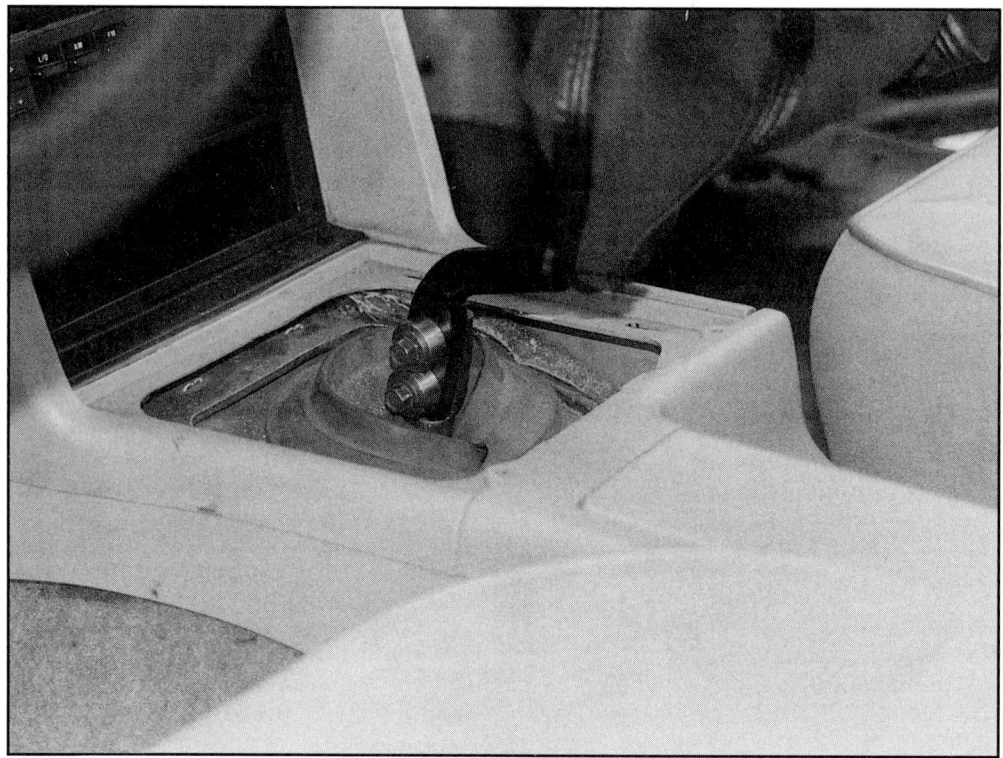

4. Once the bolts have been tightened, re-secure the shift boot.

5. With the new Steeda stainless-steel shifter bushings in place and the shift boot reinstalled, people will be hard pressed to see any difference, but you'll sure feel it.

TRANSMISSION OIL COOLER

The addition of a properly rated oil cooler for your transmission will greatly increase its life expectancy and give you years of trouble-free service. Choosing an oil cooler is one area where you don't have to be afraid of going too big, as in exhaust and camshaft selections. B&M's oil cooler design regulates the oil so that even if your cooler is too big it will only flow enough oil to cool to the correct temperature and bypass the rest. After speaking with B&M's tech people and discussing our 5.0's purpose in life, what we had done to the car, as well as what we plan to do to the car, they suggested their 24,000 GVW Super Cooler (part# 70264).

Their Super Cooler series uses a "stacked plate" design which is resistant to rocks and other road debris, as well as being able to cool more efficiently than an equivalently rated tube and fin style cooler. This allows a more efficient cooler in a smaller package. This is great for tight spot installations like the Mustang. B&M's technical department suggests an average operating temperature between 170 degrees and 220 degrees. Our stock AOD had previously seen 220 degrees in stop-and-go traffic, and racing situations, and this was before the transmission was modified! So you can see how important it is to have a cool running transmission. Any transmission shop will tell you that most repairs to transmissions and torque converter failures are due to excessive heat. B&M also carries Super Cooler engine oil coolers and remote mount transmission oil filter kits, as well as many other transmission parts like valve body modification kits and torque converters. We felt it would be the perfect time to install a transmission oil cooler since there was no bumper or nose cover to photograph around on this 5.0.

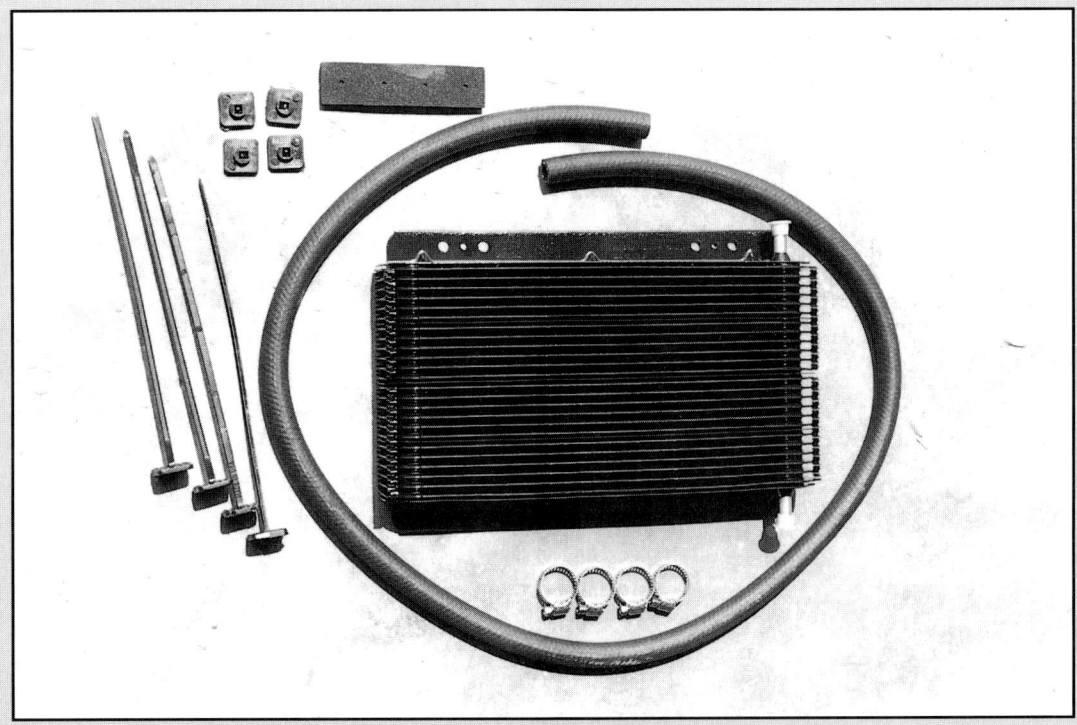

B&M supplies all coolers (except for their race only and street rod coolers) with everything you need to install your oil cooler on your daily driver.

TRANSMISSION OIL COOLER

1. Unbolt the two small retaining brackets at the bottom of the air conditioning condenser (they require a twist and pull removal).

2. Placement of the cooler is tight in the front of the Mustang. With the factory power steering cooler to the driver's side and a hood latch brace (already removed) to the passenger's side, the oil cooler should be pre-positioned beforehand to ease installation. Gently let the condenser drop down.

3. Apply the four foam pads to the back side of the oil cooler. These pads protect the air conditioning condenser from chaffing or vibration damage from the oil cooler unit.

TRANSMISSION OIL COOLER

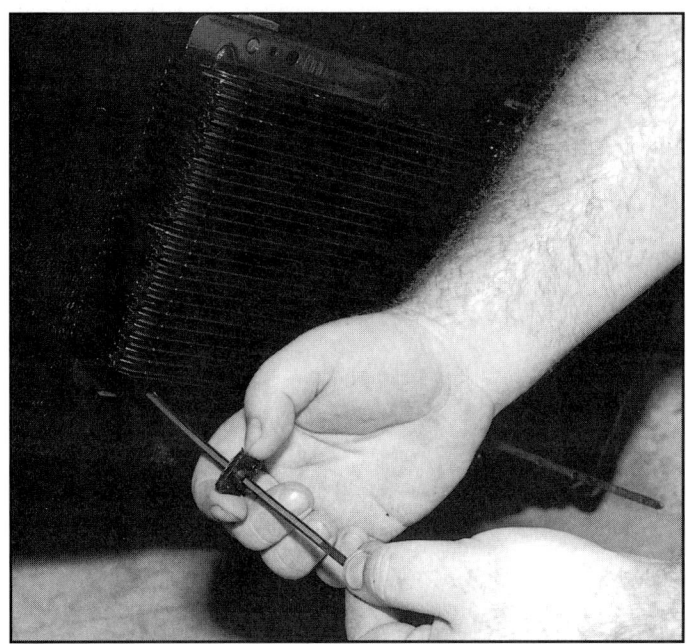

4. Using the four supplied retaining straps, pass the straps through the condenser and then through the oil cooler assembly. Slide the strap retainer over the tip of the strap and completely down until the oil cooler is tight against the condenser unit. Cut off any excess strap material.

5. Remount the condenser and attach the end of the supplied cooler hose to the bottom cooler nipple. Route the line to the cooler fitting area at the driver's side of the radiator, leaving an inch or two excess for proper fit, and cut the hose.

6. Take the remaining length of hose and connect it to the upper cooler nipple and again route it to the cooler fitting area.

TRANSMISSION OIL COOLER

7. The return line on an AOD application is the lower fitting. To check your transmission if you are not sure, follow the B&M instructions for cooler line determination. Remove the cooler line from the radiator (removing the radiator overflow bottle will give you more room to work) and install the B&M brass fitting into the radiator. Loop one hose to this fitting (arrows) and the other to the end of the cooler line (push back the fitting to allow the hose to slip over the cooler line). As you can see in this photo we also have our brass "T" junction with our transmission temperature sender installed. If you add a temperature gauge it should be installed in-line AFTER both the factory and the B&M cooler. Start the vehicle and check for leaks and check your fluid level; you might have to add half a quart or more depending upon the size of the cooler and the length of the lines.

8. Our completed cooler will fit snugly behind the factory nose assembly, but our hood latch support will have to be modified to be reused.

VI. SUSPENSION TECH

PERFORMANCE SPRINGS & SHOCKS
STRUT BRACES & SUBFRAME CONNECTORS
BUDGET BRAKE UPGRADE
PERFORMANCE LOWER CONTROL ARMS

PERFORMANCE SPRINGS & SHOCKS

At first you might think that changing all your suspension components in one weekend, or less, might sound like a daunting task, but with the right tools, and a helper, it can be accomplished in time to get you to work Monday morning.

The '87-'96 Mustangs all use the same type of suspension components, which are basically a strut type front suspension with a steering spindle attached to a lower control arm and a coil spring between the control arm and the chassis. This is the same for both sides of the front suspension. The rear is comprised of a live rear axle attached to the chassis by four control arms which have coil springs between the lower control arms and the body, and two shock absorbers mounted between the axle and the body.

Changing the suspension components for increased handling, vehicle lowering, increased tunability, or any combination thereof is nothing short of giving your 5.0 a performance transfusion. Swapping in a set of lowering springs not only makes your Mustang look better, but will handle the curves better than a stock Mustang.

Picking the right suspension parts is easy to do once you decide upon which direction you are wishing to take with your Mustang and your budget limits. If you just want to lower your Mustang for looks and have a budget of $300 you certainly don't need to be wasting your time looking at bolt-in SLA (short-long arm) front suspension systems and traction devices. On the other hand, if you are going to use your Mustang for road course or open track events, you should read up on traction devices and panhard bars, in addition to shock valving and spring selection.

Speaking of spring selection, no matter what you will be using your Mustang for, the type of spring you purchase will ultimately be your major decision factor. Coil springs come in two major types, progressive rate and linear rate. A linear rate spring is a spring that is designed to compress under a given amount of force equally throughout its length. For example, if a linear spring is rated at 600 lbs per inch, it would take six hundred pounds of pressure to compress that spring one inch, 1200 pounds for two inches, and so on. These springs offer more driver control, but can be a harsher ride.

Progressive rate springs, on the other hand, are designed to give a smooth ride at one rate, and when the second rate is achieved, the spring firms the ride up and gives the driver more control. A progressive rate spring's two spring ratings are listed together like "425/530", which means at normal compression it takes 425 lbs. of pressure to compress the spring one inch, but at full compression it takes 530 lbs. of pressure.

One option open to you, and one we recommend, would be the "suspension package" purchase option available from most vendors. Companies like Eibach, Brothers Performance (BBK), Steeda Autosports, Saleen Performance Parts, and many others, offer complete suspension systems in one kit or under separate part numbers, but designed to complement each other. These kits, or systems, usually include front and rear springs, front and rear sway bars, urethane bushing kits, front struts, and rear shocks. Most of these companies also offer upgrades or options like sub-frame connectors, auxiliary rear sway bars (installed along with the stock rear bar), adjustable shocks and struts, custom alignment specs, caster/camber adjustment plates, traction devices, and the aforementioned panhard bars.

So, as you can see, there is no right or wrong way to pick suspension parts. Though this 5.0 will be getting a combination street/drag race suspension, these instructions are nearly a mirror image for whatever type suspension you decide to use. Whether it is simply to lower your car for looks, or to eat a few Firebirds in an SCCA class, the Mustang's suspension can be made more than capable for whatever your needs may be.

PERFORMANCE SPRINGS & SHOCKS

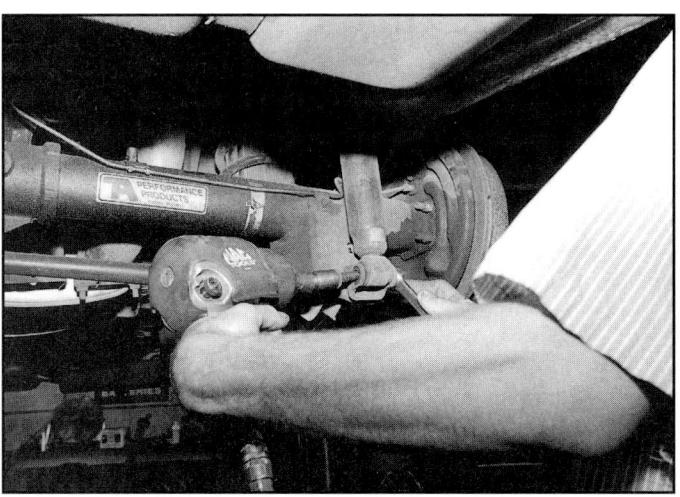

1. We started with the rear suspension by unbolting the bottom of the shocks, which allows more room to work the rear springs out. We are using a lift here for better pictures, but these same steps can be accomplished at home with a jack and jack stands.

2. Disconnecting the quad shocks from the differential will allow the differential to lower even farther, but we removed the quad shocks and brackets completely. This will let us take advantage of wider rear tires/rim offsets and will help with weight transfer for drag racing.

3. Now that the quad shocks are removed the rear springs just about fall out. A simple pop with a pry bar and they will come right out.

4. Removing the rear lower arms will require unbolting the exhaust from the H-pipe for clearance into the bolt hole in the frame for the forward arm fastener. The stock upper arms will hold the differential in place, but a floor jack or adjustable support rod should be used to prevent the axle housing from possibly dropping too far and damaging the rear flexible brake line. The rear sway bar will also have to be removed at this time. Instead of a full race piece, and to keep our budget down, we used a stock arm that had the rubber bushings melted out and new urethane bushings installed. Though we didn't show it here the Steeda instructions are thorough. You may wish to substitute an aftermarket control arm, such as a Hotchkiss, Steeda, Saleen, or Pro Mustang.

PERFORMANCE SPRINGS & SHOCKS

5. The upper arms are a simple remove and replace job. Don't tighten any suspension points yet, as you will need to load the suspension. The upper arms we decided upon are the "police" arms used on '92 and later Mustangs. These have a rubber bushing (instead of urethane) that is stiffer than the Ford Motorsport piece.

6. The Tokico Illumina Five are five-way adjustable shocks and struts to handle any driving need. The shock eye is bolted to the axle housing and then jacked into the shock tower opening.

7. The Tokico shocks require using the original upper mount washer, but the nut is spot welded to it and is the wrong size for the new shocks. Simple solution: break the nut away from the washer with a hammer and punch and install the factory washer with the two new nuts.

8. Our right rear spring houses the Poly Air bag for drag launches. The small air line is easily routed up through the spring perch, out a stamped body opening, and into the trunk/hatch via the passenger side drain plug (the one that is also most commonly used for nitrous lines). Looking back at the work done, we probably should have installed the air line onto the air bag first, instead of trying to get the hose on through the spring coils.

PERFORMANCE SPRINGS & SHOCKS

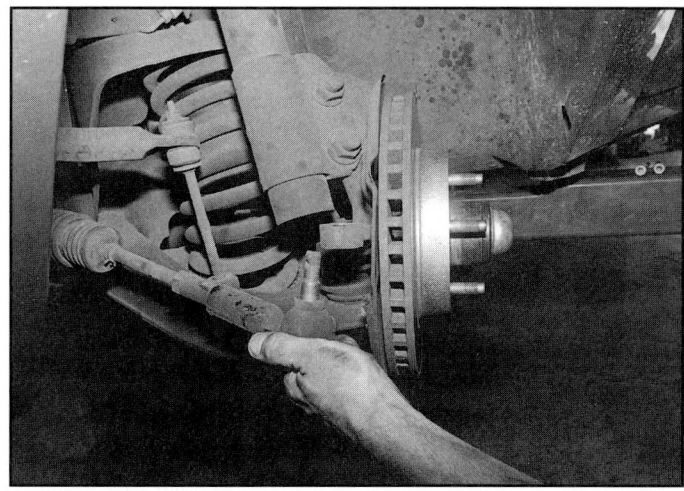

9. Remove the brake caliper and outer tie rod end from the spindle, allowing it to rotate freely from the power steering rack. The rotor and dust shield were also removed subsequently.

10. The lower arm ball joint was carefully unbolted from the spindle (the spring contains enormous pressure, BE CAREFUL AND WORK SLOWLY) with a transmission jack under the arm. The control arm pivot bolts were loosened, but not removed, and the jack was then slowly lowered until we achieved the position you see here. This allowed us to remove our tired stock springs without a compressor. If you don't have the car up in the air to a comfortable height like this, a good hydraulic jack and a spring compressor should be used for removal and peace of mind.

11. The stock front control arms were modified, like the rears, with a budget in mind. Energy Suspension carries Mustang front control arm bushing kits with bushing shells. This saves you the time and mess of burning the acrid smelling rubber from the control arm as we did for the rear lower arms. Simply press out the old bushing and press in the new urethane one (shown here) and you have it. When it comes time to reinstall the arms do not double nut them and let them swing free. This idea will ruin the bushing and shorten its life severely. While this may work for a drag only car, for the street the bushings will last longer and the car will handle better.

12. The stock strut mount and strut assembly was removed from the strut tower opening. We had a "crunching" sound whenever we turned the steering wheel or applied the service brakes; we soon found it was a bad upper strut mount that had corroded, causing a noise. Silence is golden, that's why we used liberal amounts of Energy Suspension installation grease on all urethane bushings. Jeff Tinion, of Energy Suspension, claims the biggest customer complaint is squeaking from the bushings. This all led back to improper greasing during installation. Moral of the story? Grease every new bushing as if your life depended on it!

PERFORMANCE SPRINGS & SHOCKS

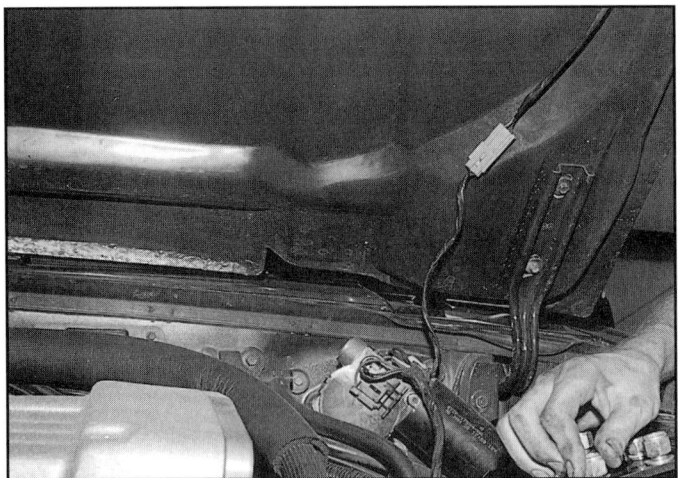

13. The adjustable caster/camber plates from Maximum Motorsports were installed in place of the stock pivot. The MM piece allows independent adjustment of both alignment specifications. If you have a '90 or newer 5.0, caster/camber plates become mandatory, as without them a lowered car can't be brought back into factory alignment specifications. Any Mustang before 1990 will have to be "bumpsteered" when the car is lowered. This is done with different tie rod ends and relocating the power steering rack.

14. The Eibach drag spring is longer than the stock spring, which necessitates a spring compressor. We tried something different here by angling the spring slightly while tightening the compressor. This points the spring in the direction of the spring saddle. Don't forget to transfer all bushings and isolators to the new springs. The front sway bar was replaced with a Cobra unit from Performance Parts Inc. with urethane bushings from Steeda.

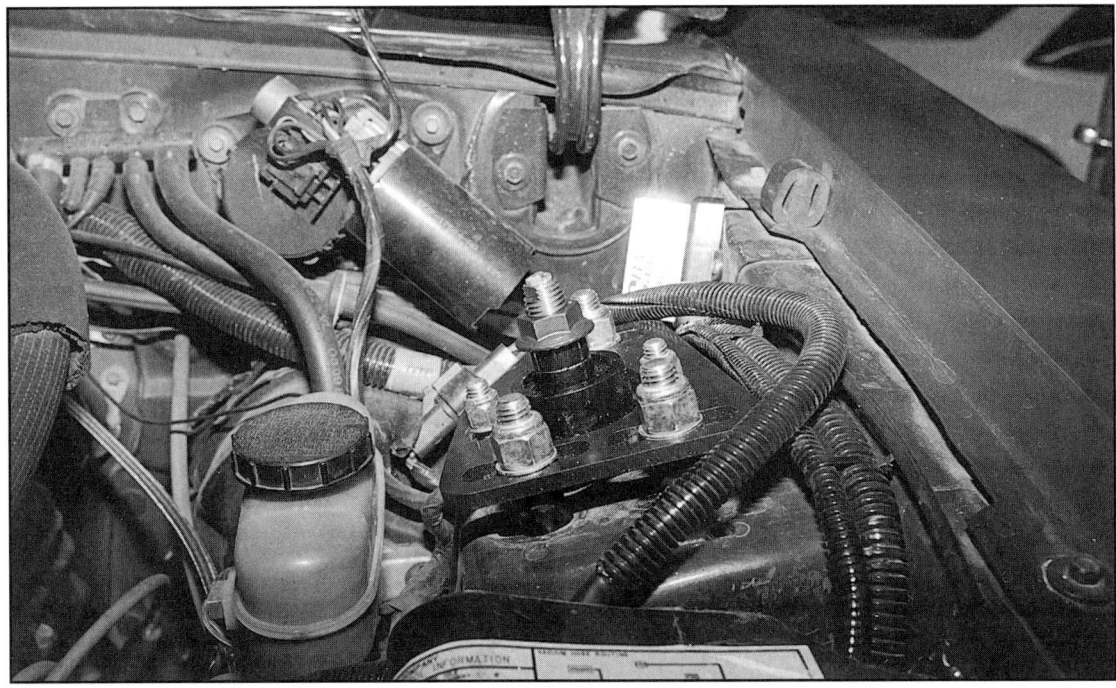

15. Once the lower arm is in place, and held by the transmission jack, the strut, spindle, dust shield, rotor, and tie rod end are all reassembled to the lower arm. Our combination of Tokico struts and Eibach springs made for the selection of shims you see here under the mounting nut. Again all applications vary, but Maximum Motorsports can help you out, as they have installed these on several different suspension setups. We were able to eyeball the toe and other front end settings to get us to a neighborhood alignment shop for the correct setup suggested by Maximum Motorsports.

PERFORMANCE SPRINGS & SHOCKS

16. The adjustment of the Tokico Illumina Five's is handled by a pocket-sized screwdriver at the top of the shock or strut. Don't take the one out of your tool box, as you will end up still needing it; go buy a new pocket screwdriver and leave it in your glove box. When we got our 5.0 back on the ground and had all the suspension pickup points tightened we set all five Tokico's on #5, which is full firm, for our test drive. Though a kidney belt might be optional equipment when at full firm, the car handled as if on rails. Several days behind the wheel will help you figure out what settings you like. For drag racing we simply unbolt the sway bar, soften the front struts, firm the rears and adjust the air pressure in the Poly Air bag for our optimum launch.

17. The TSW "Stealth" wheel in 16 x 7.5 with the correct Mustang offset were rapped with 245/50/ZR16 BFGoodrich Comp T/A ZR4 tires. This combination is what is termed a "plus one" in the wheel and tire industry. The wheel has a one inch larger diameter over the stock wheel but the tire size makes the overall height of the tire/wheel assembly the same as the stock wheel. This creates better handling by shortening the side wall of the tire. You also will not have any problems with speedometer gears when you have a "plus one" application. A "plus two" application would be a same size height with a 17" rim, such as the '94 Mustang option and the '93 Cobras.

18. The Eibach drag race springs were designed to not drop the car down as far as their regular springs, as this would hinder weight transfer at the race track. We were told we would probably see about a half inch in front and none in the rear, but we checked ride height before and after on the same surface and recorded about a half inch of drop all around. Once the front springs settle we may see a slight extra drop in front.

STRUT BRACES & SUBFRAME CONNECTORS

With the advent of unibody cars in the early '60s, the Big Three believed these cars would save manufacturing costs by not utilizing a full frame under the body of the car. This in fact did use less steel and save the auto manufacturers money, but with using the actual floor pans as a structural member of the car, any floor pan damage or deterioration would effect the structural rigidity of the vehicle. Many early Mustang owners know of the heartache and hassle that the replacing of the old rotted floor pans can cause. Convertible owners for years have complained of sagging cars and doors that are hard to open and close.

One thing that can be done to solidify these unibody cars is to "tie" the subframes together with sub-frame connectors. Other techniques include strut tower braces, which tie the front strut towers to the firewall for reduced flex in the front subframe area, and lower cross braces in either two or four point mounting configurations. The lower cross braces are used in conjunction with the strut tower brace to "cage" in the engine compartment area.

We obtained a complete "chassis kit" from Kenny Brown Performance Parts. Their chassis kit consists of a three point strut tower brace, a lower four point cross brace (sometimes called a K-brace), and their double-cross subframe connectors. This complete chassis stiffening kit has been extensively tested on late model convertibles and has shown an improvement of over 70 percent in the reduction of static torsional twisting.

Now you don't have to have a convertible to feel the improvements that this system can make to your car; any coupe or hatchback can benefit from this system by improving weight transfer and improving torsional rigidity throughout your vehicle. Whether you drag race, like to cut corners, or just want to reduce unibody seam separation and floor pan fatigue, these chassis parts will work for your needs. Kenny Brown Performance Parts also sells various roll cages and other suspension enhancements, including their AGS Systems (Advanced Geometry Suspension), which take the Mustang from a potent street runner to a Mustang that will beat a Porsche or other high dollar exotic car in the corners as well as the straights.

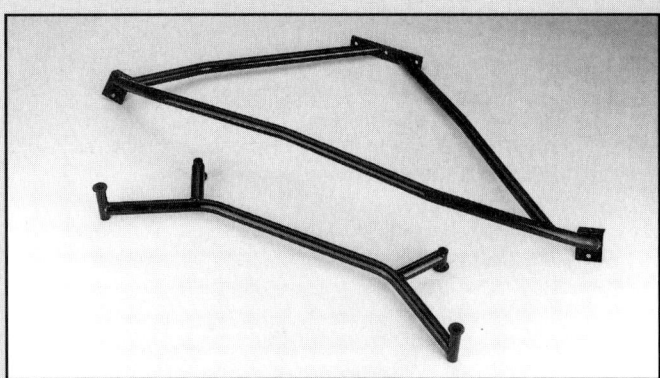 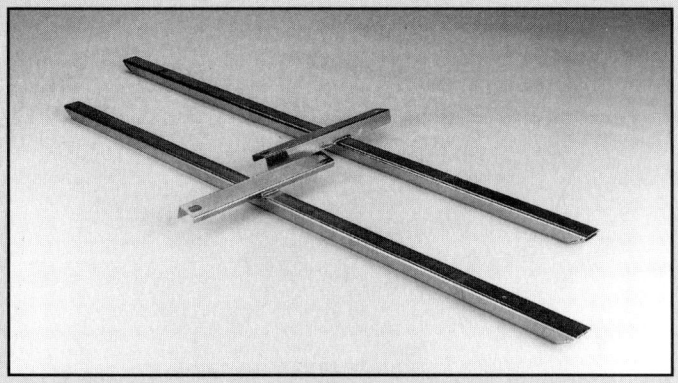

At the left is the Kenny Brown three point strut tower brace and the lower four point chassis brace in standard black powder coat; they are both available chrome plated. At the right is their double-cross subframe connectors in the new zinc chromate plating which resists rust.

STRUT BRACES & SUBFRAME CONNECTORS

1. Our task of installing these components requires only a drill and some basic hand tools, save for the sub-frame connectors which are a weld in only affair. If you already have bolt in sub-frame connectors we strongly urge you to weld them in place. The bolts never stay tight enough to let the connectors work to their full advantage. The three point strut tower brace will be our starting point for this addition to our 5.0. Kenny Brown carries several models of strut braces for supercharged vehicles as well as GT-40, Edelbrock, and Cobra intakes. Remove the MAF meter mounting bracket bolts (if equipped) and move the meter out of the way. Also remove the coil cover at this time.

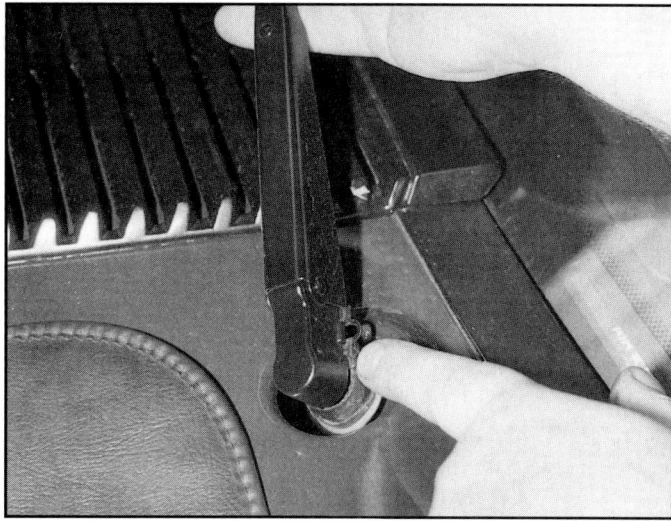

2. The wiper arm assemblies and the cowl grille will both need to be removed to gain access to the cowl area to install the mounting bolts. To remove the wiper arms, simply pull the arms up gently until they rest against their stops and then pull the locking pin, located at the base of the arm, outwards (finger). While holding the locking pin outwards, let the wiper arm return to the windshield. The wiper arm should now be resting approximately two inches from the windshield. If it is resting correctly, as described, the wiper arm can now be lifted free of the transmission arm shaft.

3. Six screws are all that retains the cowl grille to the vehicle. Once the six screws are removed, disconnect the washer hose from the washer nozzle and remove the grill from the car. You might want to wax the painted surfaces before reinstalling the grille (since you don't get under it too often).

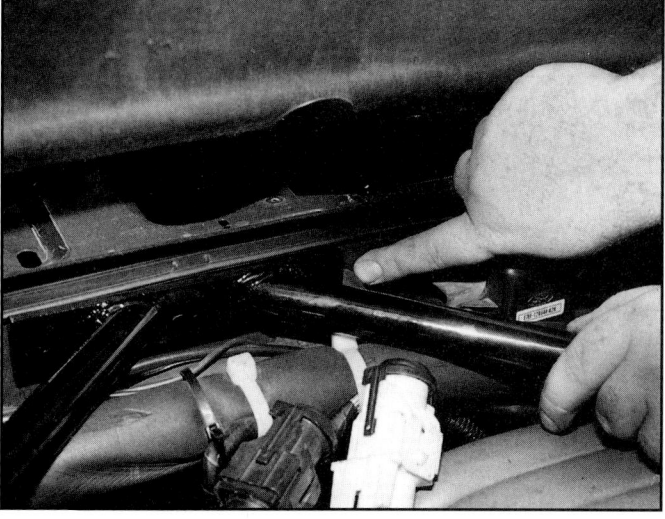

4. Place the strut tower brace in position on the strut towers and ensure that the firewall mounting pad is flush against the firewall and butted directly to the right of the washer nozzle hose where it enters the firewall (finger). Mark the two outboard mounting holes on the firewall with a punch or scratch all and remove the brace from the car.

STRUT BRACES & SUBFRAME CONNECTORS

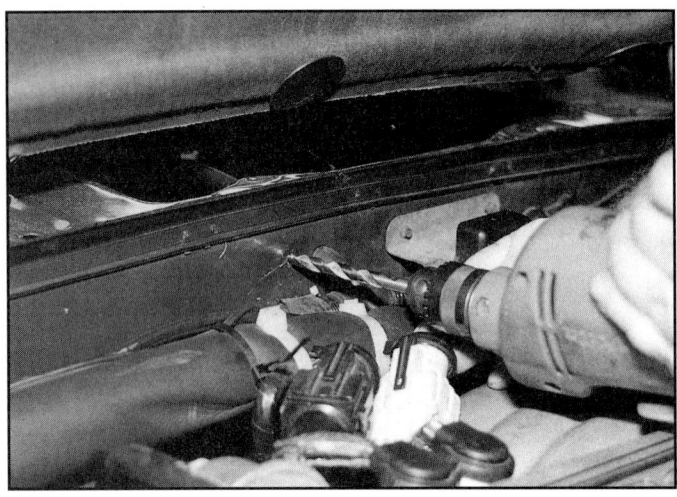

5. Drill the two previously marked locations with a 5/16" drill bit.

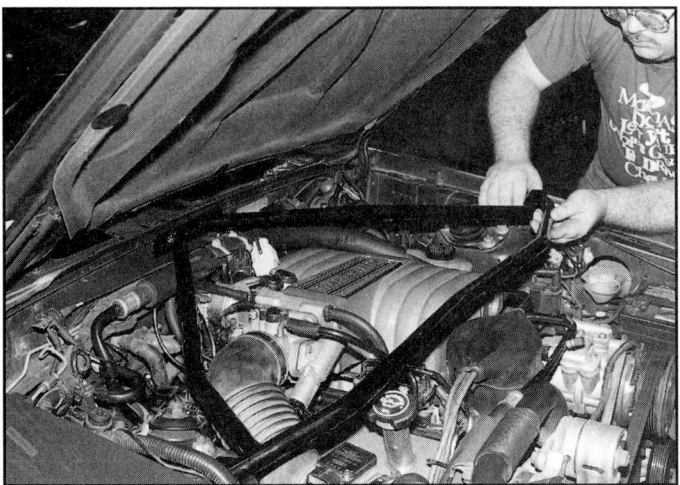

6. Place the strut tower brace back in place and insert the two outer mounting bolts through the firewall and into the brace with the supplied washers and secure with the lock nuts per the included instructions. Once secured, drill the center hole and install the supplied mounting hardware.

7. We had to cut this corner off to allow the MAF meter bracket bolt to be reinstalled in the same hole. This was the only installation glitch of the whole kit.

8. With the strut tower brace semi-secured in the car, drill out the four mounting holes at the strut towers.

STRUT BRACES & SUBFRAME CONNECTORS

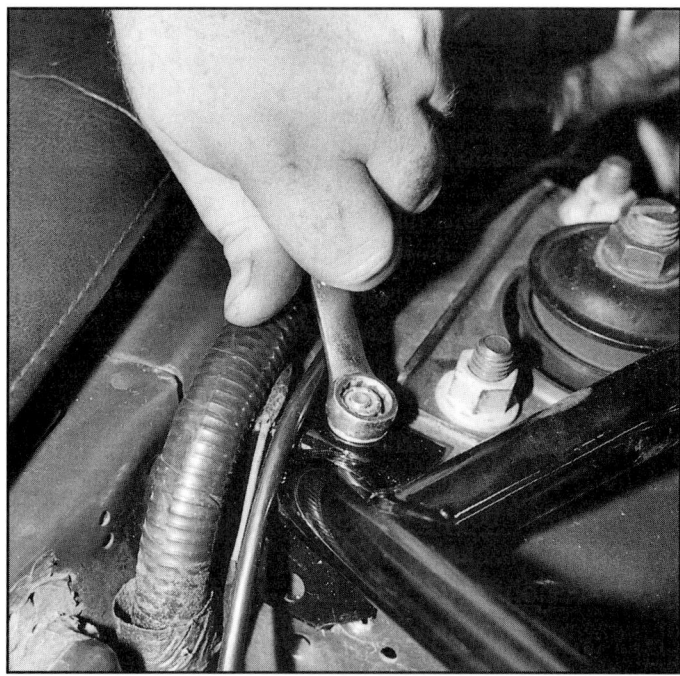

9. Install the supplied mounting hardware to complete the installation. You may benefit by jacking up the car momentarily to reach the drilled holes in the strut tower.

10. Reinstall the MAF meter bracket and the coil cover. The cover can be carefully placed below the strut brace (as shown), or the cover can be trimmed with a razor knife to fit around the strut brace.

11. Our completed strut tower brace looks right at home under the hood of this 5.0. Though the powder coating looks great, you may wish to have the brace painted your car's body color for that "factory" look.

12. Now we go on to the double-cross sub-frame connectors. These connectors use a cross brace that cuts the free span of the connector in half by bolting the connector to the strongest part of the floor pan area, which is the front seat rear mounting bolts. Here, the sub-frame connectors are held in place while the weld points on the sub-frames are marked with a pencil.

STRUT BRACES & SUBFRAME CONNECTORS

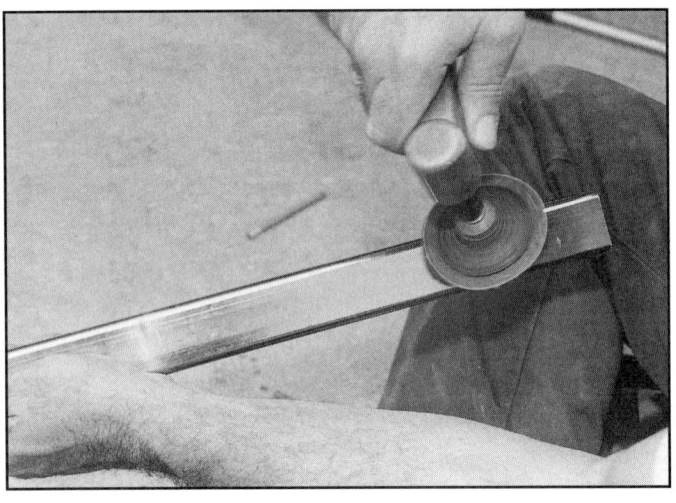

13. Once the areas to be welded are marked, the factory sub-frame areas and the edges of the connectors are cleaned of all paint and chemicals to promote the best welding adhesion.

14. The connectors are then loosely installed with the supplied hardware. Once the sub-frame connectors are positioned parallel with the factory cross members, the mounting bolts are tightened completely and welding can begin.

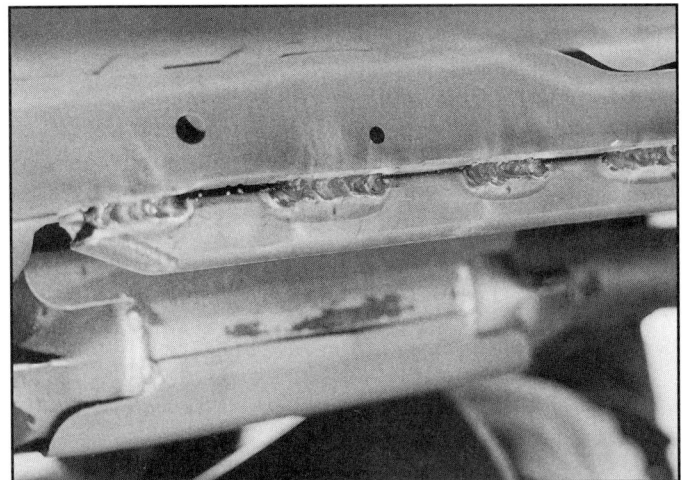

15. The sub-frame connectors are welded on both inboard and outboard sides with two inch stitches. Complete stitches along the entire mounting length are not really necessary, but you may wish to take the time and completely weld the entire mounting length. Finish the sub-frame connector installation by spraying the welded and unprotected areas with a rust neutralizer and then some rust proof paint or light undercoating.

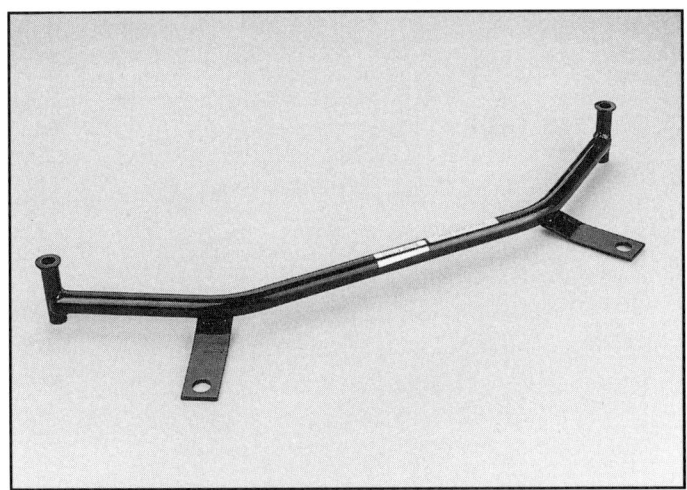

16. This 5.0 had long tube headers, which occupies the same space that the lower four point brace would usually use. In this case a two point brace would fit, but not give us the torsional rigidity of the four point mounting. The brace you see here was designed by Kenny Brown Performance Parts for cars using long tube headers. The instructions included are excellent and the lower brace bolts right in with the four mounting bolts included.

BUDGET BRAKE UPGRADE

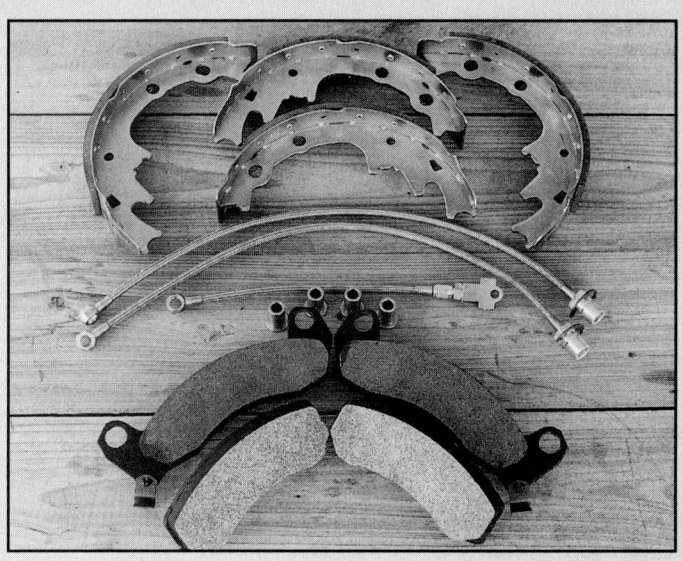

Maximum Motorsports brake components, all available separately, include Hawk Racing street compound front disc pads, premium quality rear linings on new shoes, their own braided stainless steel Teflon lined flex hoses, and stainless steel caliper guide sleeves.

Today's high performance vehicles (and some not so high performance vehicles) are shod with large front disc brakes, as well as rear disc brakes in most cases, but '87-'93 FOX Mustangs use 10.8" front disc brakes with nine inch rear drum brakes. While this poor excuse for a "performance" brake system can be upgraded substantially by brake experts such as Saleen, Wilwood, Baer Racing, Aerospace Components, and Brembo, with 12" or 13" front disc brakes along with 12" rear disc brakes to make your Mustang stop as fast as a Trans Am Mustang, there are some major disadvantages to these systems. One is that they can severely lighten your wallet, and two, wear items like disc pads have to be ordered from the brake manufacturer. Also, some of these systems require permanent modifications to the spindle or other part of the front suspension. While these systems are excellent for SCCA Mustangs or serious street/strip Mustang owners, a modest, strong acting street system is not out of most people's budgets.

There are several ways to approach upgrading your brakes within your budget. You can upgrade your rear drum brakes to disc brakes with any one of several brake kits. You can replace your front disc brake calipers with those from the Lincoln Mark VII, which offers a larger piston diameter, while maintaining the stock exterior dimensions of the caliper. You may also replace your stock pads and shoes with performance pads and shoes, and install disc caliper guide sleeves and stainless steel braided brake lines, or any combination of these. While the rear disc brake kit is the most expensive of these budget approaches, getting rid of the stock nine-inch rear drums puts you that much more ahead of the game in stopping power.

Actually installing the brake components is a matter of basic mechanical knowledge and common sense. If you don't want to worry about having to "bleed" brake lines, stick with replacement parts like performance pads and shoes only. If you don't have a fear of opening the brake system up then by all means install the larger front calipers and braided brake lines. The most labor intensive job mentioned so far would have to be the rear disc brake kit, as the differential must be drained of oil and the rear axles removed to install the disc brake caliper mounting brackets. The Mustang also uses a bevy of different fasteners, from Torx ™, to US, and metric sizes. In addition, some brake lines are metric and other lines are US, thus a well-stocked tool box should be considered to prevent multiple trips to the tool store.

Maximum Motorsports budget performance brake components for late model Mustangs can be installed easily in your driveway and don't alter the stock component's layout (caliper and rotor size) in any way. These new brake parts will increase your Mustang's stopping power dramatically with no major modifications. While replacement performance pads must still be ordered through a vendor, stock replacement pads can be used in an emergency.

BUDGET BRAKE UPGRADE

1. Removal of the front disc caliper requires the removal of the two guide pin bolts from the caliper assembly. Once these guide pin bolts have been removed, the caliper can be pulled free of the rotor and spindle assembly.

2. Without the proper tool for removal, the careful use of a screwdriver can remove the dust cap from the rotor. Insert the tip of the screwdriver between the cap and the rotor and gently twist the screwdriver. Repeat this procedure all around the cap until it is removed.

3. Once the dust cap is removed, remove the cotter pin to access the spindle nut. Proceed to remove the spindle nut retainer, spindle nut, washer, and outer bearing as seen here.

4. With a grease pencil mark both left and right rotors for proper reassembly.

BUDGET BRAKE UPGRADE

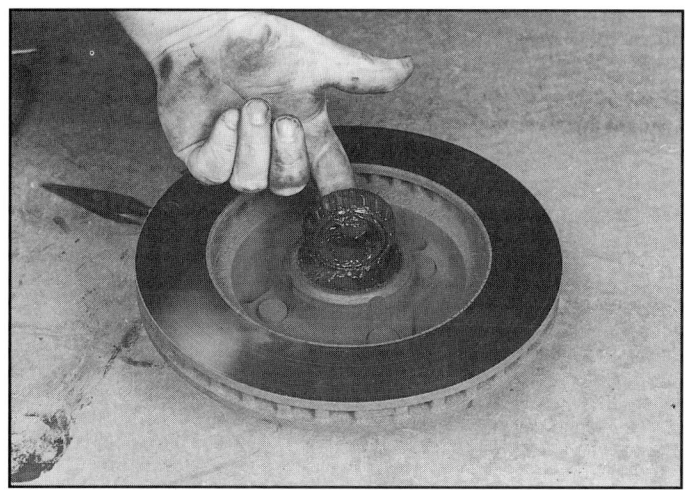

5. Once the rotor has been removed from the car, remove the grease seal and the inner bearing. Keep all left side bearings together and all right side bearings together.

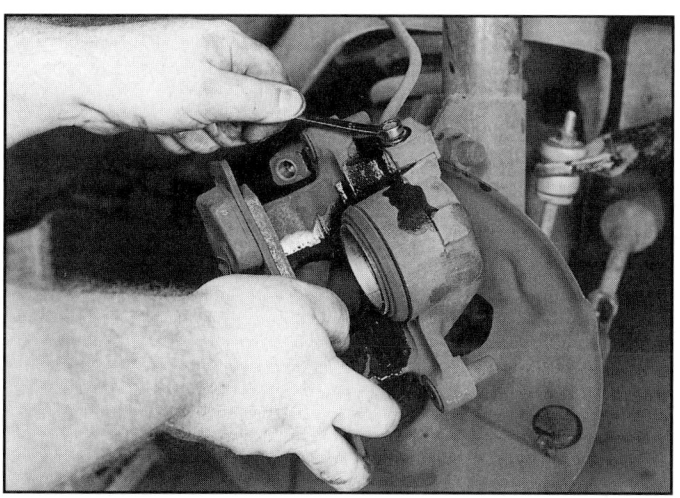

6. Using a 10mm boxed end wrench or 10mm socket, remove the banjo bolt at the rear of the caliper. Once the caliper is free of the flexible brake hose, place the caliper on the bench to ready the caliper for its brake upgrades. Ensure that the two copper washers located on either side of the banjo bolt fitting at the end of the flexible brake hose stay with the hose, they will not be reused.

7. Use a line wrench or "tubing" wrench to break free the brake line fitting at the top of the spring perch.

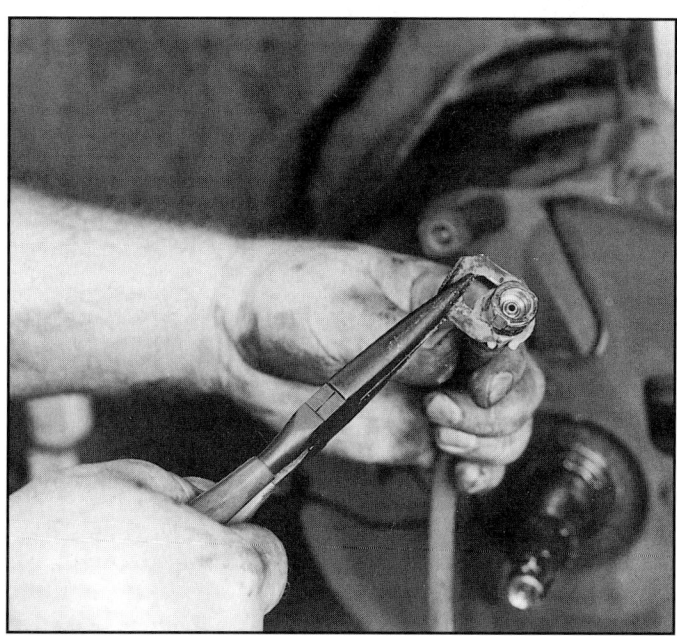

8. Once the hard brake line has been disconnected from the flexible brake hose, remove the flexible brake hose by removing the horseshoe shaped retaining clip from the retaining groove on the hose.

BUDGET BRAKE UPGRADE

9. With a 10mm socket, remove the retaining bolt for the flexible brake hose retaining bracket.

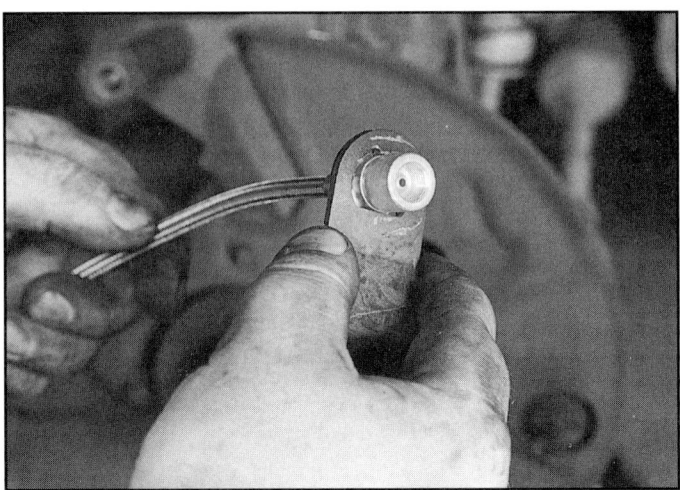

10. File the small retaining tab, located on the retaining bracket, just enough for the new braided brake hose to pass through it.

11. With the old inboard brake pad still in place, compress the piston with either a large C-clamp or a large pair of adjustable pliers. Compress the piston slowly and evenly to prevent binding of the piston in the piston bore. Inspect the piston for any cracks or torn dust boots. Any damage of this nature will require a rebuild of the caliper or the purchase of a rebuilt unit. One option is upgrading to 73mm piston calipers, available through most 5.0 performance outlets, or from any new parts vendor as a Lincoln Mark VII OEM caliper.

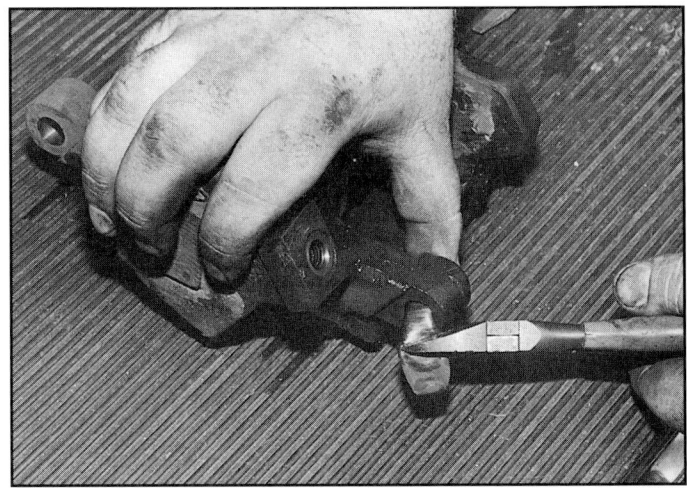

12. Remove the rubber caliper guide sleeves from the caliper. If you plan to save yours, care should be taken during removal.

BUDGET BRAKE UPGRADE

13. The instructions state that the aluminum caliper guide sleeves should be pressed into the caliper, but a small section of wood and steady even blows of a ball peen hammer will suffice in getting the job done.

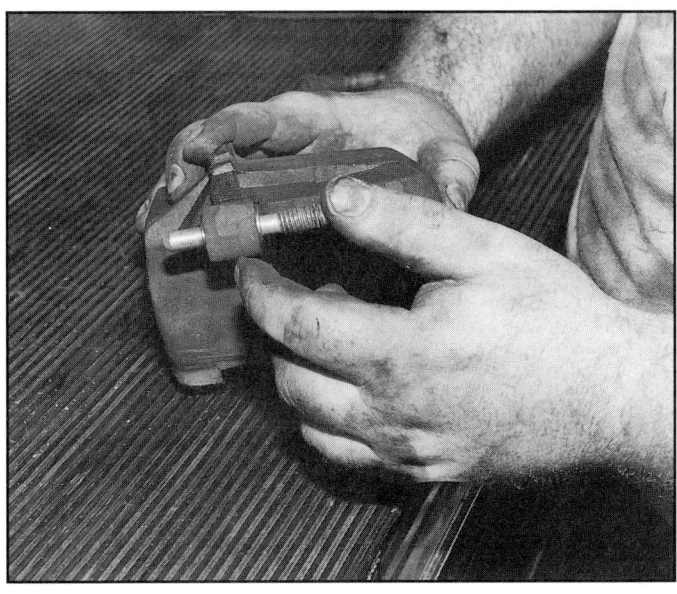

14. Once the aluminum sleeves have been installed in the caliper, insert the guide pin bolts and check for free movement. If the guide pin bolts bind in the new aluminum guide sleeves, run an 11/32 drill bit through the guide sleeves to clean them up.

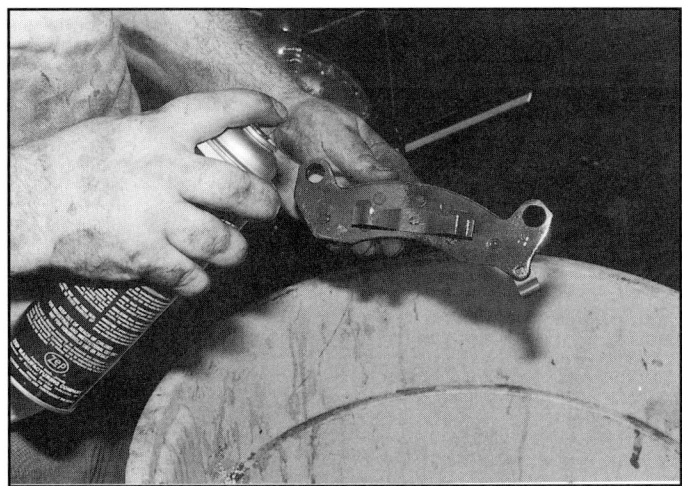

15. A good spray-on or brush-on anti-squeak solution will go a long way towards keeping your brakes quiet.

16. For optimum braking efficiency Maximum Motorsports recommends having the rotors lathe turned, which will require a trip to your local auto parts store or jobber. Once our rotors had been turned we proceeded to repack the wheel bearings and install a new grease seal. Again, another option would be to cross drill your factory rotors, or upgrade to a stock sized aftermarket rotor, such as the ball milled and cadmium plated GTRotor.

BUDGET BRAKE UPGRADE

17. Re-install the rotor and the outer bearing, along with the washer and spindle nut. Tighten the spindle nut to 17-25 LB-FT while rotating the rotor and then back off the spindle nut one half turn and then re-tighten to 16-28 LB-IN (finger tight).

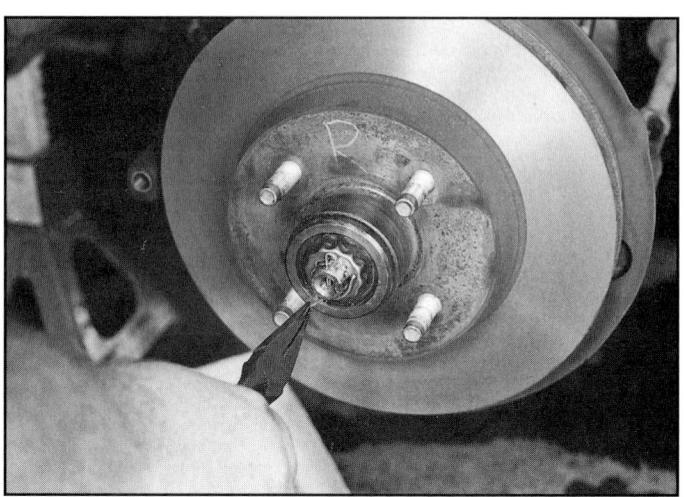

18. Install the spindle nut retainer in such a manner that the new cotter pin can be inserted through the hole in the tip of the spindle. Bend back the tips of the cotter pin to retain the rotor assembly and install the dust cap.

19. Re-install the caliper to the spindle bracket followed by the braided steel line connected to the caliper. Ensure a leak free connection by using the new copper washers supplied with the brake hoses. Position the brake hose away from the caliper and towards the inner fender at a 45° angle to prevent tire rub. Add the supplied tie wrap to keep the brake line out of harm's way.

20. Remove the rear flexible brake hose from the axle assembly by removing the single retaining nut located on the lower left of the axle housing. This will also necessitate the removal of the left and right rear brake line fittings at the hose junction. Mount the rear hose junction block in a vise and either file or grind away the edge of the junction block to facilitate removal of the bracket from the hose.

BUDGET BRAKE UPGRADE

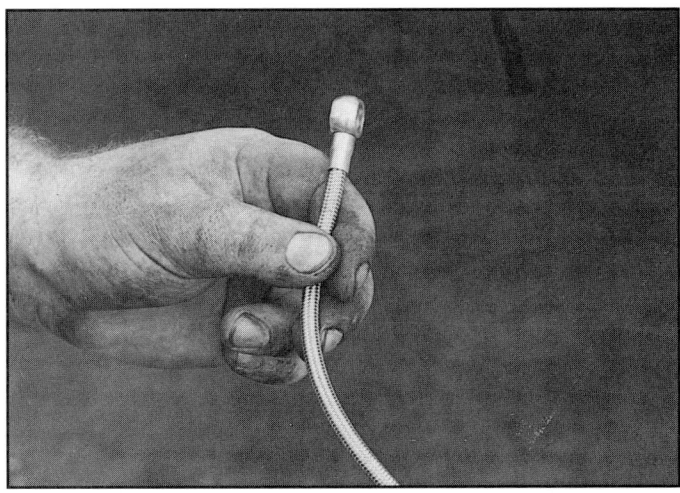

21. Using a small file or a drill, enlarge the opening to allow the new retaining bolt to pass through the bracket. Mount the new braided hose and junction block to the original bracket as shown.

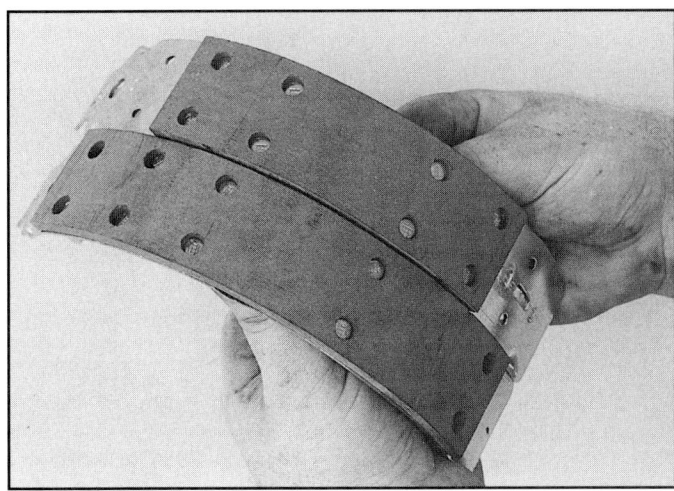

22. The brake shoe linings sold by Maximum Motorsports are made of all new materials. These premium quality linings are riveted to new backing plates instead of bonded like the factory pieces. The brake shoe with the shorter lining is called the primary shoe while the remaining shoe with the longer lining is called the secondary shoe.

23. Remove the brake drum and clean the entire brake area with brake cleaning spray. This type of cleaning is more effective than the older type of vacuum cleaning.

24. Drum brakes can be a bit more complicated than front disc brakes. It would be wise to invest in a brake handbook or a shop manual that covers the braking system in question. The tool shown here is used for the removal of the primary and secondary springs. To use the tool place the head of the tool over the spring post and rotate the tool so that the tool's ear will rotate around the spring and free the spring from the shoe.

BUDGET BRAKE UPGRADE

25. The secondary spring will be removed first, then the primary spring. The primary spring is always re-installed first.

26. The shoe retaining spring (shown here already removed) uses a retaining pin that is positioned through the backing plate then through the brake shoe and retained by the spring. To disassemble the spring, hold the retaining pin with your fingertip and apply pressure to the spring head. With the spring still under pressure, rotate the spring until the head of the pin is positioned in such a manner to allow the spring to disengage the pin and decompress.

27. With the drum brakes completely disassembled, the backing plate is cleaned and inspected for any serious signs of wear.

BUDGET BRAKE UPGRADE

28. In preparation for the new brake shoes, some high temperature grease is applied to the backing plates where they support the brake shoes.

29. Reinstall the parking brake bracket to the secondary shoe and install the shoe to the backing plate with the retaining spring.

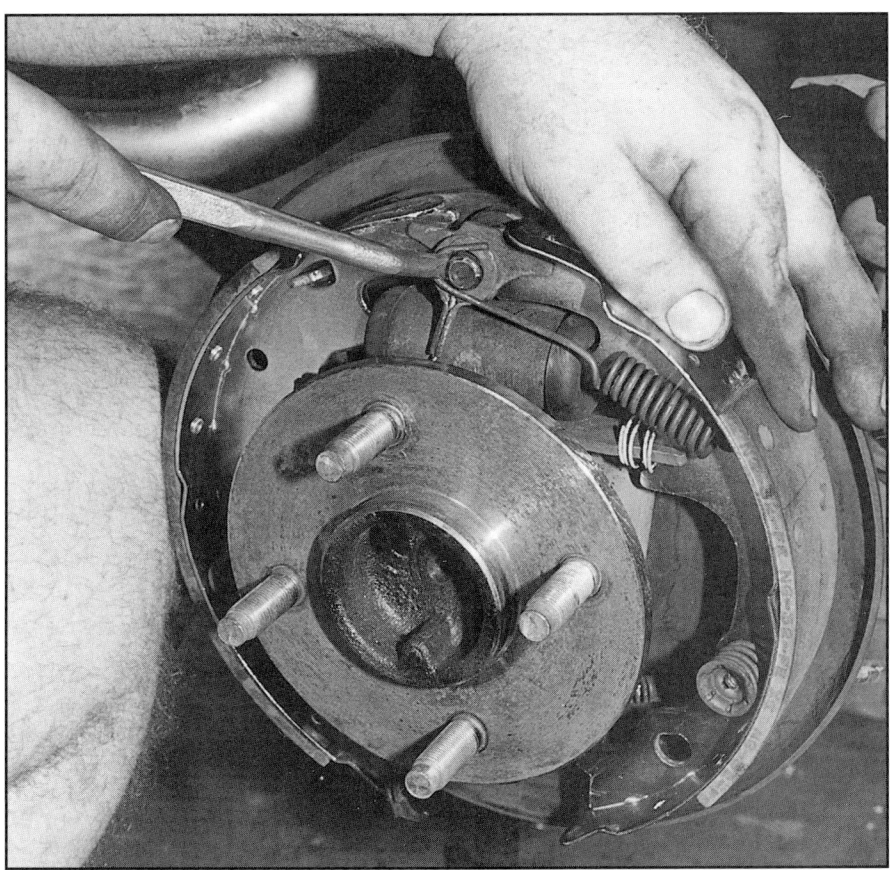

30. Install the primary shoe and parking brake link onto the backing plate. Install the shoe retainer, self adjuster cable and primary and secondary springs. Here the primary spring is being installed. The primary spring is always installed first.

BUDGET BRAKE UPGRADE

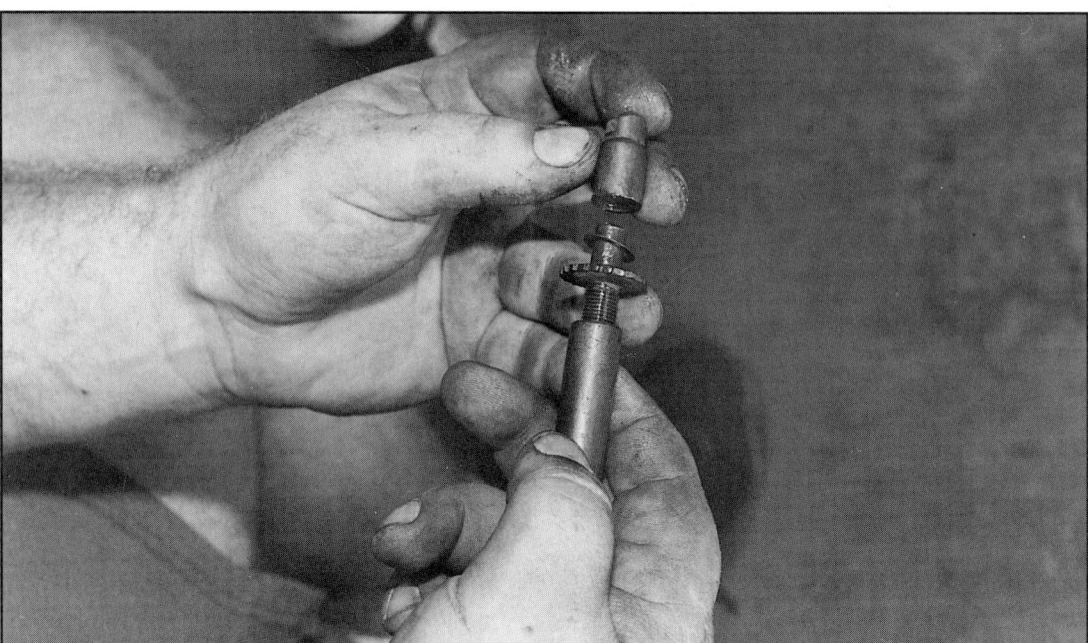

31. Disassemble and clean the self adjuster mechanism. Do not lose the small washer during cleaning. Lightly grease the threads and the washer and reinstall the self adjuster assembly to the brake shoes, along with the self adjuster lever and spring. Again it is very wise to refer to a brake handbook or shop manual for drum brake repairs. It should be noted that the passenger side adjuster is left hand threaded.

32. With the rear brakes completely assembled, turn the self adjuster (it should click with each détente of the adjuster star wheel) until the brake drum just slides over the shoes for optimum rear braking efficiency. Follow the instructions included in the Maximum Motor sports kit for proper brake bleeding. Maximum Motorsports recommends either Ford or Castrol GT brake fluid for the best in braking effort. They recommend staying away from DOT 5 silicone fluid for the Mustang system.

PERFORMANCE LOWER CONTROL ARMS

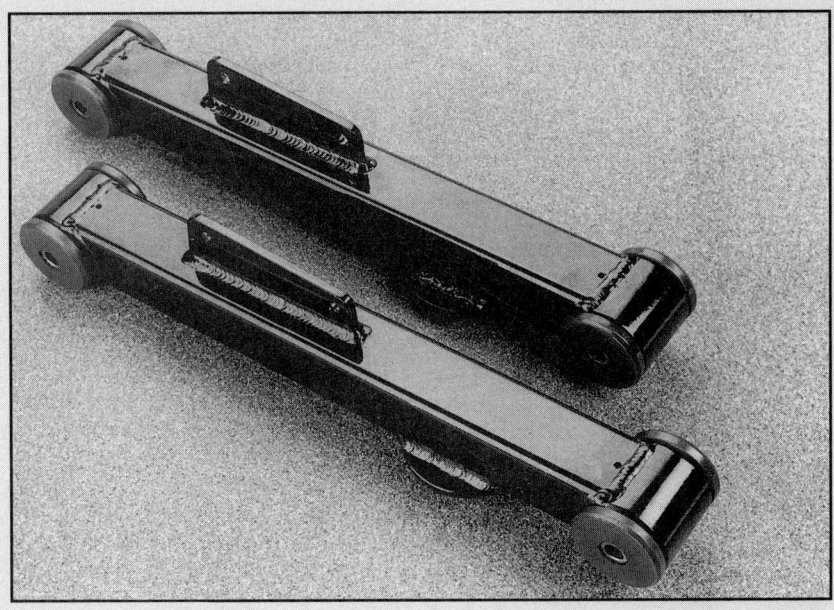

Steeda Autosports' new lightweight aluminum control arms may not look like much, but in this case looks are definitely deceiving. The control arms bolt in place of the stock arms, retaining the rear sway bar, with the use of metric tools and a floor jack.

While we just showed you how to completely replace all your suspension components for increased performance and handling, if you are on a part by part budget or can only earmark part of your dollars for suspension build-up you can't go wrong with installing just rear lower control arms.

With a large selection of rear control arms available, it is difficult, if not impossible, to figure out which one you will be happy with. Several types of control arms are available, including drag and road race versions, adjustable arms to control pinion angle, aluminum, steel, rubber bushings, urethane bushings, and some even have no bushings at all.

Steeda, after many hours of testing other control arms, developed these control arms for the 5.0 Mustang that will improve handling and turn-in while offering a real world ride. So what is so different about them? For starters, two different types of bushings are used in the same ends of the control arm. The center bushing carries a 95 durometer rating, and the outer bushings are an 80 durometer in their road race versions and 88 in their drag race versions. For an example, the stock arm with rubber bushings has a durometer rating of just 50-66. In conversations with Steeda's New Product Development Coordinator, Dan Carlson, we were told the road race bar will work great for the typical street 5.0 of 300 horsepower and lower, but when the fire is raised to the big numbers that can produce 1.50 60-ft. times and run 12.00 ETs, the drag race bar should be used. The bushings are available in kit form to change the bar from one type to the other easily.

The arms themselves are made of lightweight aluminum for a measurable weight savings over the stock steel arms. The stock arms with urethane bushings in our project car weighed in at seven pounds, while the Steeda aluminum arms with bushings weigh in at a scant 4.3 lbs. That may not sound like much, but the percentage of weight (especially unsprung weight) is good. These control arms are also much lighter than steel versions on the market that use extra brackets. Steeda's new control arms are made from boxed material for greater strength over the stock open stamped arm.

PERFORMANCE LOWER CONTROL ARMS

1. Raise the rear of your Mustang to a comfortable working height, place the body on jack stands, and remove the rear wheels. We will be tackling the installation one side at a time, so the following captions will not list a specific side; just pick a side and get started. Begin by unbolting the rear sway bar and set it aside for re-installation later.

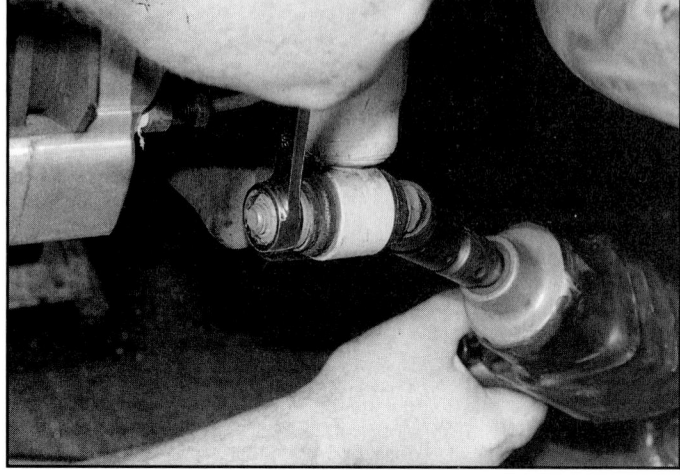

2. Slightly raise the rear axle to take the axle's weight off of the rear shocks and remove the lower bolt. A wiggling of the axle assembly or shock will help get the bolt out.

3. Using an 18mm socket and an 18mm wrench, remove the nut, but not the bolt, from the rearmost attaching bolt of the lower control arm.

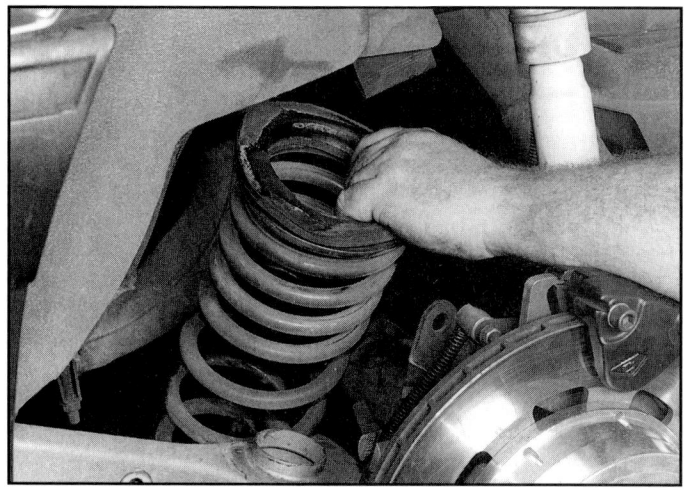

4. Place a floor jack under the control arm and jack up the arm until the bolt can be pulled free of the axle and control arm. Slowly lower the jack until you can remove the coil spring from the car. Set the spring and the upper isolator aside. The lower isolator can be ditched, as Steeda's instructions say it isn't reused.

PERFORMANCE LOWER CONTROL ARMS

5. The forward mounting bolt of the control arm is in the rear frame rail of the Mustang. This bolt and nut are also 18mm. Depending upon your exhaust system you may need to loosen the muffler/tailpipe assembly to get to the bolt head inside the frame. Once the bolt is out, walk the control arm out of the frame section.

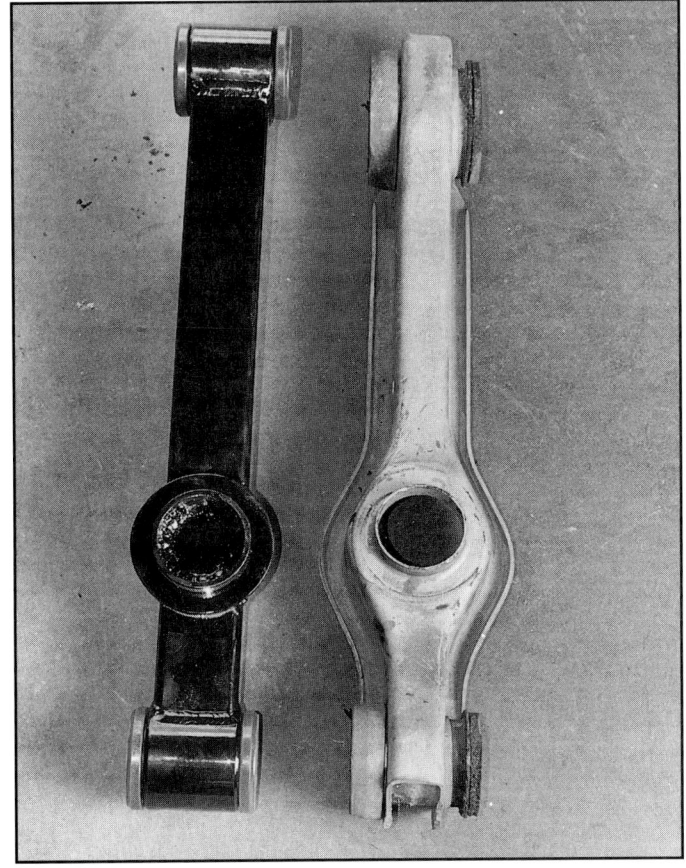

6. The stock arm is shown here next to the Steeda lower arm. Since the Steeda arm is boxed for strength it doesn't need to have all the material and curves like the stock arm. This allows for deeper offset rear rims and wider rubber. Another advantage was that we were ditching the stock oval bushing in the front of the stock control arm, which gives more deflection than a round bushing. The lower unsprung weight offered by the light aluminum arms is another advantage over the stock arms.

7. The outer bushings have flutes cut into them to retain grease and come pre-greased and assembled from Steeda, thus no grease fitting is required, or its associated maintenance.

PERFORMANCE LOWER CONTROL ARMS

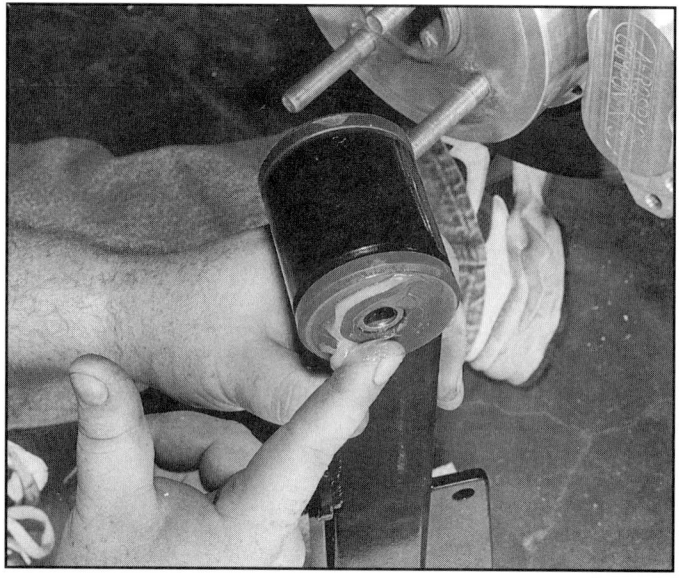

8. While the new Steeda lower control arms come pre-assembled, the sides of the bushings will need to be greased to prevent squeaking. We always recommend the sticky marine grease that resists water, for obvious reasons. Any marine supply store can help you out.

9. The installation of the arms is the reverse of removal. Simply install the front bolt loosely, position the spring with the end towards the rear of the car, and raise the arm with a jack and a piece of wood. Another jack may be needed to change the pinion angle for the holes to line up. We used a small bottle jack to lift the front of the axle housing.

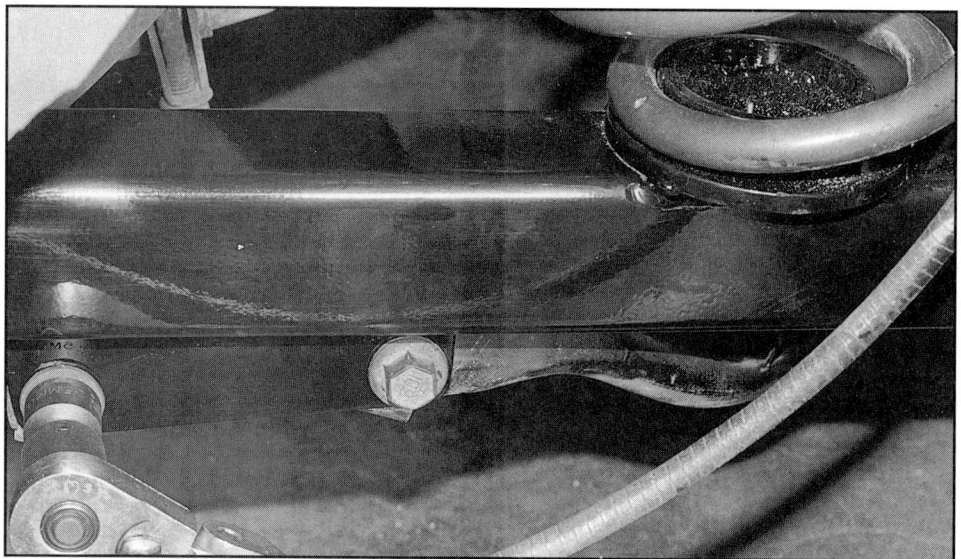

10. The provision for the stock sway bar requires turning the sway bar's threaded clips around to bolt the sway bar in place. The sway bar is also placed on the outer side of the control arm, not the inner as shown here. If you place it on the inner side of one arm the other arm won't reach the sway bar. The sway bar location is moved back a few inches to offer greater clearance for the popular aluminum rear end covers on the market. The relocation of the sway bar makes it act as a larger sway bar, which in turn increases handling and offers quicker turn-in.

VII. INTERIOR TECH

STEERING WHEEL/SHIFT KNOB REPLACEMENT
BODY SUPPORT BAR
AUXILIARY DASH GAUGES
REMOTE FUEL DOOR RELEASE

STEERING WHEEL/SHIFT KNOB REPLACEMENT

Our Momo leather wrapped wheel was complemented by one of their latest designs in shift knobs. The Mustang cruise control hub and cruise control horn pad are separate items.

When most people think of bolt-on enhancements their mind immediately wanders to headers, blowers, wheels, ground effects, and so on. There is nothing really wrong with that, but have you looked inside your Mustang lately? Aren't you getting tired of grabbing that 29 cent plastic shift knob and that rubber dipped steering wheel?

Take a good look at most of the Mustangs on the road. They have nice paint, nice wheels, some body modifications here and there, and usually enough power and/or stereo equipment to propel them to the outer stratosphere, but what about the interior? Most Mustang owners simply tint their windows so you won't see the stock interior. Something that you can do about that "seen it a million times" interior is to spruce it up with a new steering wheel and shift knob.

Adding a new steering wheel not only enhances your stock interior, but can offer improved driver control and feel of what your Mustang's chassis is doing. Often times replacing the steering wheel means having to lose the convenience of the OE steering wheel mounted cruise control switches. Not so today. Many vendors now incorporate aftermarket switches, or allow you to relocate the stock switches into the new steering wheel or into a mounting plate between the steering wheel and the steering column.

There are dozens of replacement shift knobs for five speed manual transmissions on the market. Some use the stock shifter handle's threads for attachment while others use a triangular setting of set screws to retain the knob and keep it from moving. Still others use threaded inserts to allow one type or style of knob to fit several applications simply by inserting the correct threaded adapter. The latter two of these three designs allow for a larger selection of handles to fit a broad cross section of applications. As in steering wheel selection, the choices are limited only by your taste and checkbook.

Momo USA, one of the world's largest makers of aftermarket steering wheels, shift knobs, and wheels has the answer. No minor finish work or modifications are needed and each one of their wheels, whether it is leather wrapped or wood finished, is hand made by skilled craftsmen. With so many wheel and shifter styles to choose from, including colored leather in fuchsia, red, blue, yellow, green, and black, there is bound to be the perfect wheel and shifter combination for your Mustang's interior, and your wallet. Our anatomic leather wrapped wheel in black will complement our stock black dash area, plus we added an anatomic grip shifter to keep our hands planted where they belong in the twisties.

STEERING WHEEL/SHIFT KNOB REPLACEMENT

1. Before starting make sure your front wheels are straight and the steering wheel is centered. The factory horn pad is removed by firmly pulling the horn pad towards you. It would be wise to disconnect the battery before starting, as you will be working with live wiring (the horn circuit).

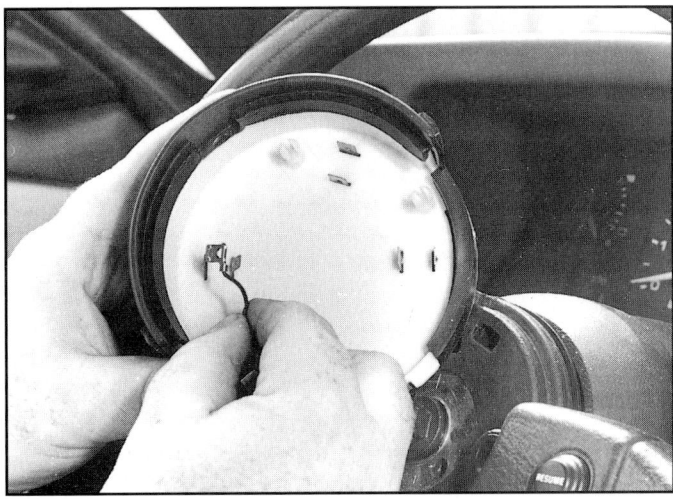

2. Disconnect the two wires from the horn pad back and place the horn pad aside. There is no need to mark these wires as the horn button is nothing more than a momentary contact switch connecting the two wires.

3. Using a long ratchet handle for leverage, remove the steering wheel mounting bolt. If the steering column is "locked" it will give you more leverage to loosen the mounting bolt.

4. Install a steering wheel puller to the wheel and remove the wheel from the steering shaft.

STEERING WHEEL/SHIFT KNOB REPLACEMENT

5. Remove the seven screws that hold the cruise control/horn brush contact ring and steering wheel back plate and remove the two switch assemblies and contact ring from the old wheel. You are now finished with the old wheel. Place the screws in their original holes for storage and reinstall the horn pad. The wheel can now be stored for future reinstallation.

6. Disconnect the three terminal plug from the contact plate and feed the wiring through the steering wheel hub adapter as shown.

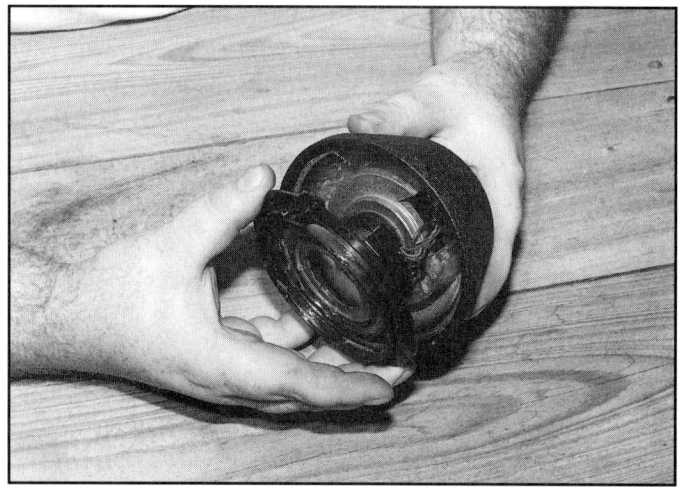

7. Reconnect the wiring plug to the contact ring and press the ring into the hub adapter until it seats flush. You may wish to clean and lubricate the brushes (on the steering column) and the contact ring for better cruise control operation.

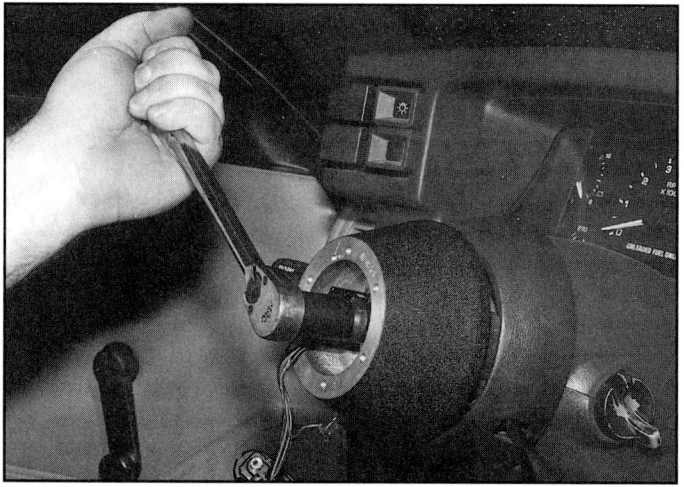

8. With the wires carefully out of harm's way, install the adapter hub to the steering column using the factory retaining bolt.

STEERING WHEEL/SHIFT KNOB REPLACEMENT

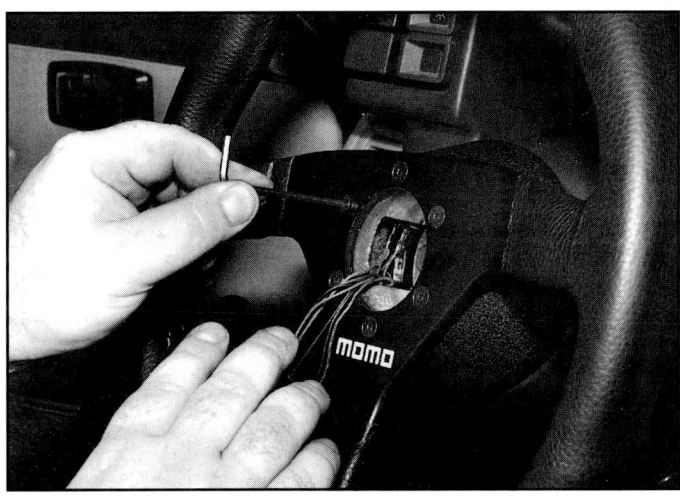

9. Install the Momo leather wheel to the adapter with the supplied Allen head screws and Allen wrench. Tighten them evenly in a criss-cross pattern.

10. Place the Momo supplied horn button into the center of the horn pad and install the two horn wires to the contacts on the back.

11. Press the cruise control buttons through the horn pad and then seat them from the front of the pad. Install the pad over the three spokes of the Momo wheel and you are done. Reconnect the battery and check for proper cruise control and horn operation.

12. Remove the factory shift knob by simply unscrewing the knob counter-clockwise until it is free from the threads. Push the shift boot down as much as possible for room to work.

STEERING WHEEL/SHIFT KNOB REPLACEMENT

13. Slide the Momo shift knob base ring over the shift handle and let it rest there until you are finished.

14. Install the proper rubber spacer for the top of the shift handle. Our Hurst linkage, with stock knob threads, took the second largest rubber spacer. This keeps the handle centered on the shaft since there are no threads inside the handle. This is due to its universal application. The base is held by three set screws while the rubber top keeps the knob centered.

15. Install the three set screws, either the long or the short, depending upon the diameter of the shifter shaft. We used the long set screws for a secure fit on our shifter. Orient the shifter to the way you would like it to feel while shifting and tighten the set screws with the supplied Allen wrench.

16. Slide the base ring, installed in step 13, up to the base of the shifter and screw it on snugly. Pull the shift boot up the base of the shifter at this time to complete your shifter installation.

BODY SUPPORT BAR

When it comes to interior/exterior body appearance enhancements it seems that the latest rage is to make your Mustang into a "full on" looking race car. Many people are adding "roll" bars, gauges on the cowl panel in front of the windshield, big tires out back and skinny ones in front, and other such race trim to set their car apart from the rest of the other half a million Mustangs produced since 1979.

While most of these race type items are simple bolt-ons, interior bars for the most part use to mean stripping the interior to the bare metal and leaving your Mustang in a racing fabrication shop to have a "cage" bent up and welded in place. This was very expensive, time consuming, and not worth the effort for a street car when you are only going for the looks. After a while vendors saw the need for a "universal" interior bar for the people that are shopping in the "looks" department. The problem with these bars is that being universal means extra time to fit the bar to your application and they usually require cutting trim panels to fit.

With the third-generation Mustang becoming such a big aftermarket parts hit, it was just a matter of time before the universal bars gave way to direct fit units. We called upon Dugan Racing for some insight on the many different types of interior bars available, such as the single hoop, four-point, and six-point designs they sell for coupes, convertibles, and hatchbacks, and at that moment they were readying the installation of one of their six-point bars into an '89 GT hatchback. Follow along as we peek over their shoulders through the installation.

Dugan's six point interior bar is made of .120 wall thickness seamless steel tubing that is mandrel bent to fit the interior. No trim panel cutting is needed, though on hatchback cars the hatch cover pull shade is rendered unusable due to the rear bars. Though this bar has been pre-painted in aqua to match the trim of the black GT, Dugan's interior bars come shipped in raw steel for the purchaser to finish. This allows for the purchaser to choose from painting, powder coating, chrome plating, etc. for their bar. For the weekend do-it-yourself crowd, spray cans will give you a durable finish for the time invested.

BODY SUPPORT BAR

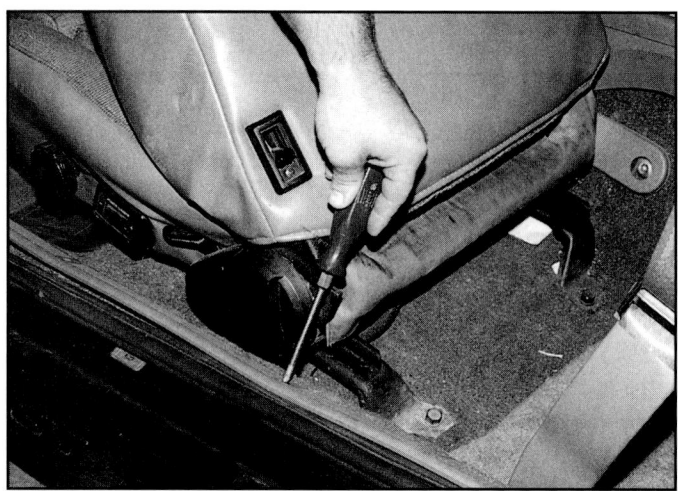

1. Very minor trim removal is needed to "mock" the bar up. Tom Dugan recommends that you make sure your bar fits your car correctly before it is painted. They will not take returns on modified or painted interior bars. Remove the four screws that retain the door sill plate and remove the sill plate.

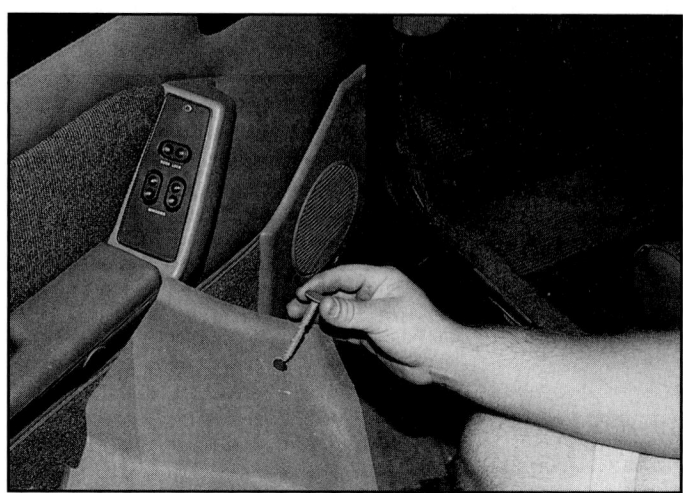

2. With the sill plate removed the forward kick panel can be removed simply by removing the plastic retaining pin at the front of the panel and then sliding the panel out. Repeat steps one and two for the passenger side.

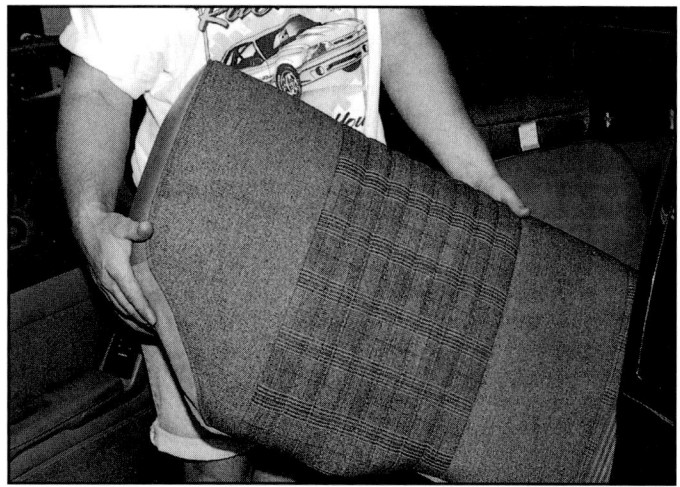

3. Push rearward on the rear seat cushion and lift up to disengage the cushion from the floor hooks and remove the seat cushion.

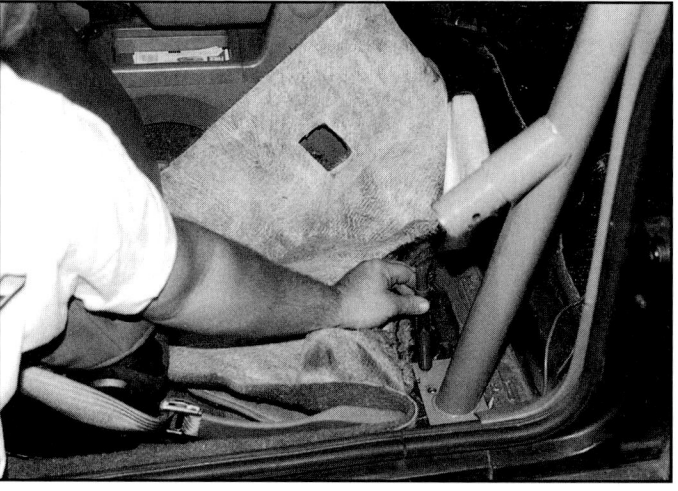

4. Pull back the edge of the carpeting in each corner and position the main hoop of the interior bar in place and have at least one other person hold it (two would be better) while you mark the eight holes for drilling with a permanent magic marker.

BODY SUPPORT BAR

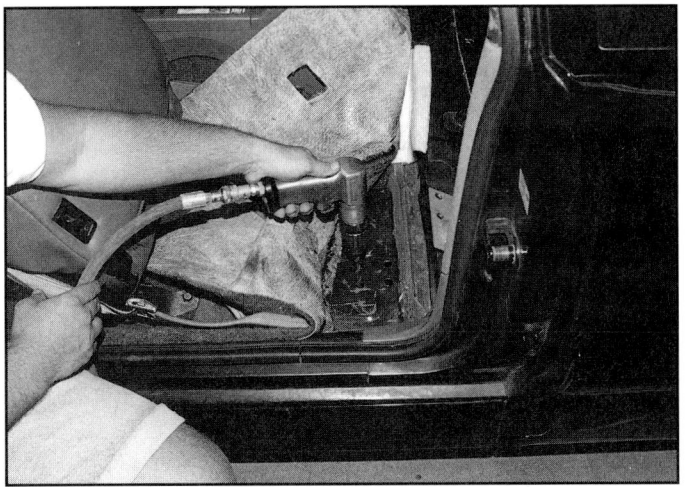

5. Temporarily remove the hoop and center punch the marked areas and drill the eight holes with a 3/8 inch drill bit.

6. Reinstall the main hoop and bolt it in place. Don't forget the reinforcement plate under the car (which should be painted or undercoated) and use a small amount of body sealer around the mounting holes to prevent water entry or rust.

7. Insert the rear mounting bars into the main hoop and ensure the mounting plate is flat with the trunk floor before tightening the rear bar to main hoop bolts. Drill out the eight mounting holes for the rear bars.

8. Install the long bolts for the rear bars with the reinforcement plates, which straddle the rear subframe, for added strength.

BODY SUPPORT BAR

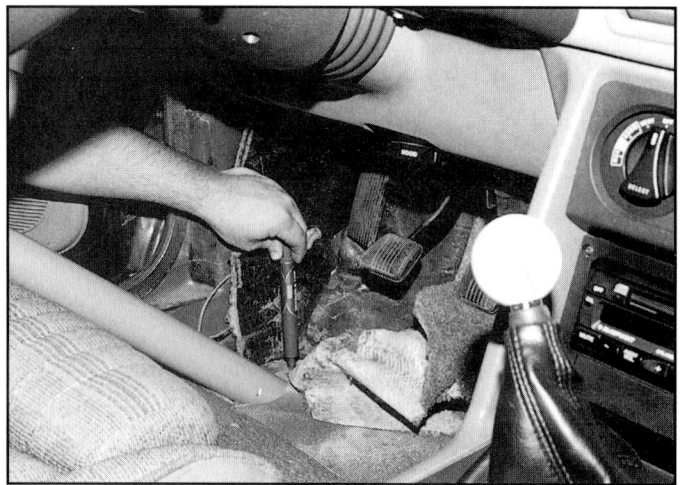

9. Insert the forward interior bars into the main hoop and tighten the mounting bolts, ensuring that the forward mount is flush with the front floor pan, then mark the holes for drilling with a magic marker.

10. Center punch the marked areas and drill your mounting holes. If you have trouble getting your drill into this area you can drill the holes from the underside or use a right angle drill, available from most rental stores. Install the mounting bolts along with the reinforcement plates. Reinstall the rear seat cushion and all the trim panels previously removed.

11. The front bar allows easy access to the front seats. There is no need to climb through a jungle gym to get inside. This bolt-in interior support bar can be used for some race applications, but check with your sanctioning body for rulings. Some require a cross bar behind the front seats. Dugan sells this cross bar, but it is weld in only.

12. The rear bars allow plenty of storage area access and room for speakers, nitrous bottles, etc. The seats can still fold down and lock in place with no problems when using this bar.

AUXILIARY DASH GAUGES

One look at a 1987 and newer dash and you would be hard pressed to find a nice clean location for a set of aftermarket gauges. The factory gauges, especially the oil pressure, are nothing more than electro-magnetic analog gauges that usually read inaccurately. In fact, the 2.3L oil pressure gauge uses an oil pressure switch that "turns" the gauge on when oil pressure is above 7.5 psi. The gauge does not actually measure pressure like the 5.0 system. This inaccuracy can lead to engine devastation when not properly monitored.

We're sure that, like yourself, there are many others that simply ignore their factory gauges, cursing the oil pressure gauge as it falls to the red zone while cruising. One answer has been the Gauge Cage. While an excellent choice for the race crowd, more of us street minded types in the hotter climes (such as Nevada, Florida, and Arizona) would like to be able to install a set of gauges and keep our center outlet a/c vents.

This is not a manufacturer's kit, but a home grown idea that we thought we should share with you. The gauge location at the forward end of the console, just beneath the stereo (common sense tells us that if you have the factory equalizer or aftermarket stereo equipment, then you will not be able to mount your gauges as we show here) makes for a clean installation while maintaining full visibility of your gauges and effortless full A/C cruising. The only owner fabricated part will be the gauge face plate. We have listed full dimensions below. The more popular 2 5/8" gauge will not fit in this area, but the slightly smaller 2" and 2 1/8" gauges will fit perfectly without any major reduction in readability.

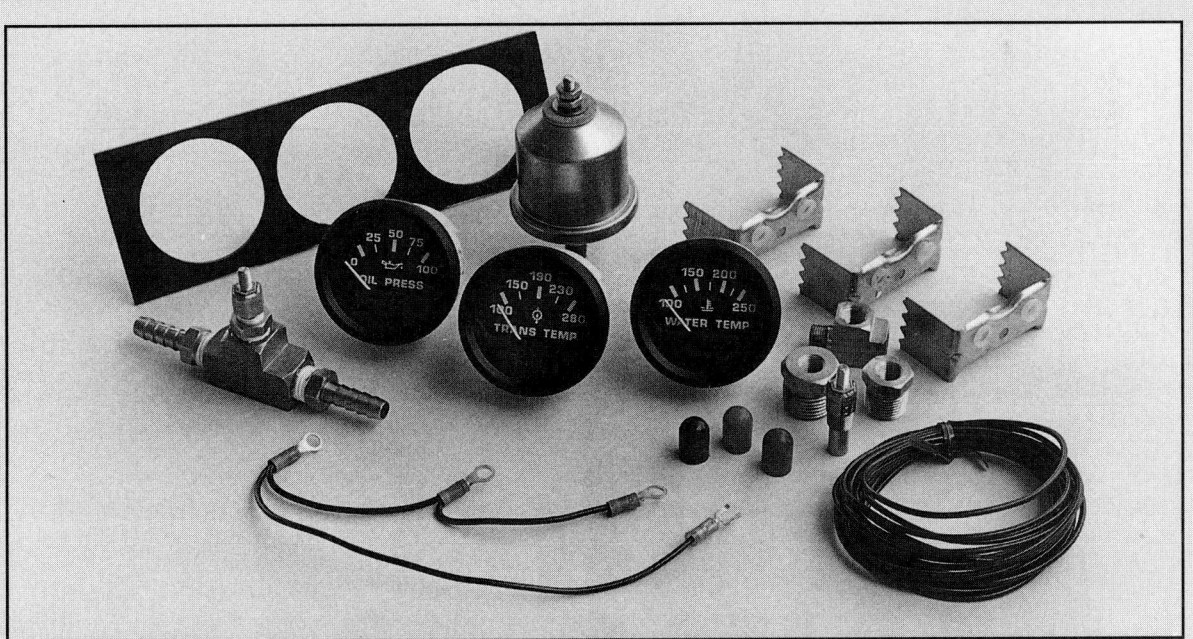

Most gauge manufacturers carry the size gauges you will need. We used three 2" gauges from Faria Performance instruments. These gauges come complete with all senders and adapters. We suggest at least an oil pressure and water temperature gauge, while the third gauge can be a voltmeter, boost, transmission temperature, etc. The face plate was owner fabricated (see side bar page 156) and the wiring kit and colored bulb covers are extra.

AUXILIARY DASH GAUGES

1. We opted for the Faria Professional Black series in this 5.0. These make a bold statement with all black plastic construction, non-glare ABS lenses, and white graphics. They are a very close match for the factory gauges and make for a clean and functional look for any late model dash. Once you have decided the order in which you wish to view the gauges, install them into your fabricated face plate.

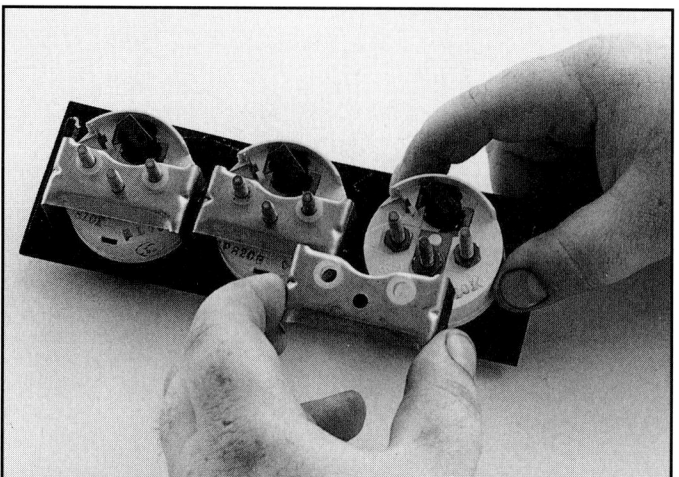

2. Our gauges were all electric for ease of installation. The Professional Black series can be ordered in mechanical versions if desired. Once the gauges are in place, slip the lock plates over the gauge terminal posts. On some mechanical gauges these posts do not carry current and don't require the plastic isolators, so if you are using a combination of mechanical and electrical, ensure you keep the correct lock plate with the corresponding gauge to prevent gauge damage.

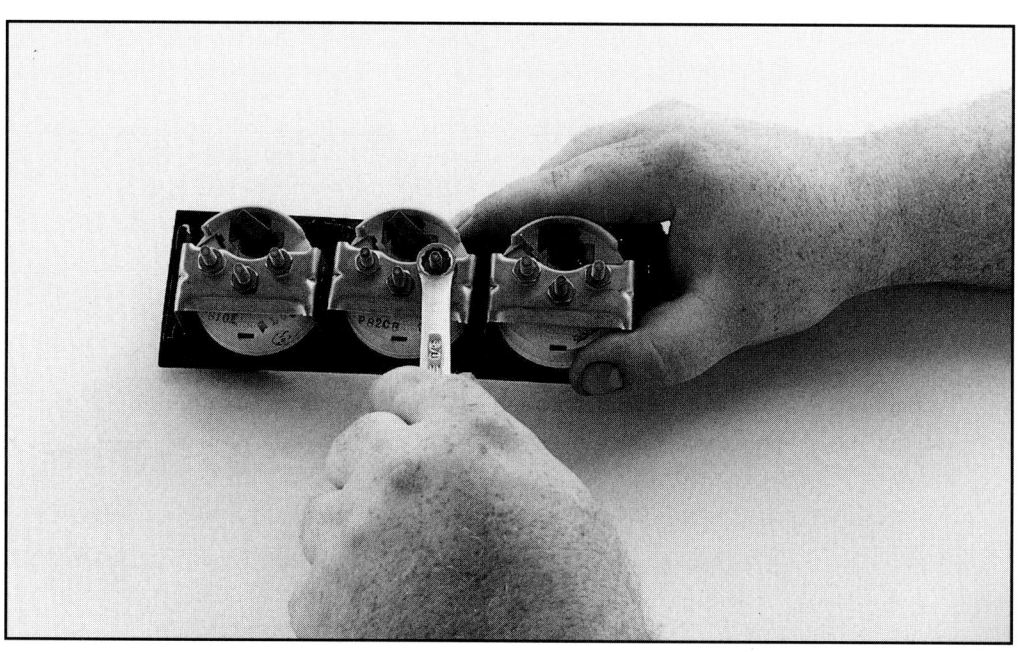

3. Once you have your lock plates properly aligned (double check the face to ensure that the gauges are in the direction you wish to view them in), install the three nuts per gauge and tighten to six inch pounds, barely more than finger tight.

AUXILIARY DASH GAUGES

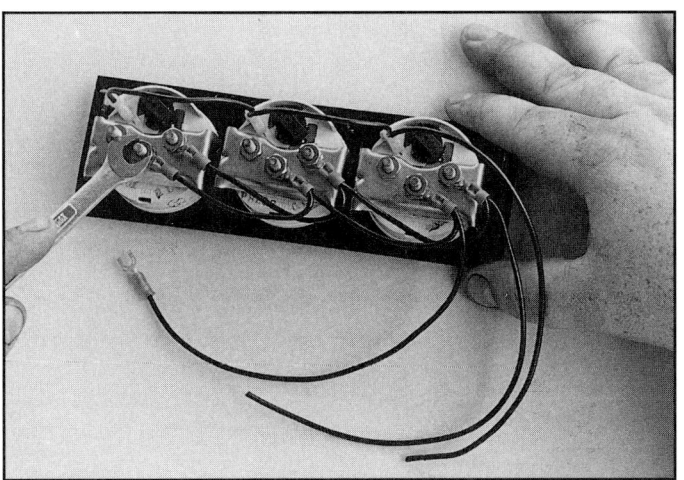

4. The wiring can be made easier by connecting the gauge's power, ground, and light connections in series. This not only saves wiring, but makes for a much cleaner installation. Here the power (right of center stud), ground (center stud), and light (upper left terminal) connections have all been made into one main wire per source coming off of the gauge panel. The ground wire has a horseshoe terminal installed on its end for a close proximity dash screw, while the other two wires will be tapped into power and ground circuits at the back of the stereo when removed and require no terminals.

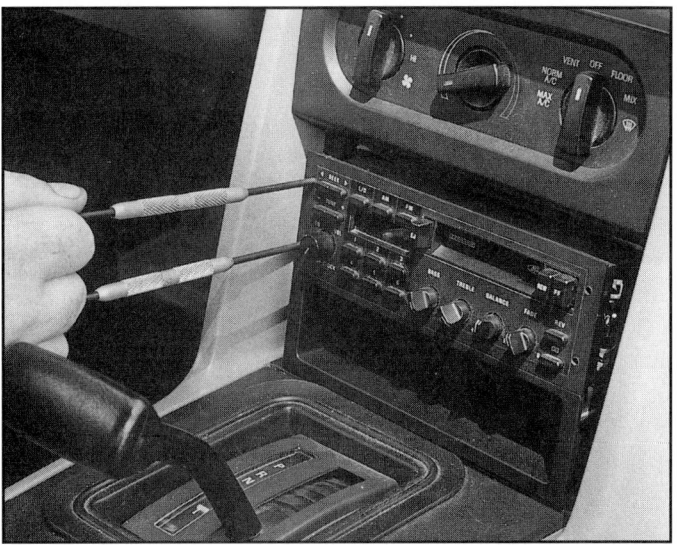

5. The stereo is removed by releasing the four lock tabs located behind the radio face plate. To release these tabs a stiff object such as a pick or small nail will need to be inserted into each of the four release holes and gently pulled outwards to compress the spring inwards. While maintaining this pressure the stereo can be slid outwards past the locks and then completely removed. This will take some patience and you may have to have a helper keep the stereo from popping back in place as you work from side to side. If you aren't comfortable with performing this removal operation yourself an area stereo or electronics shop, as well as any Ford dealer, has the correct tool and knowledge to release the stereo locks in a matter of seconds, more than likely at no charge.

6. There is no real need to disconnect the stereo electrically. Simply turn the stereo over and remove the two screws which retain the small tray to the stereo bracket.

AUXILIARY DASH GAUGES

7. Once the two retaining screws are removed the tray can be dislodged from the retaining bosses of the stereo by sliding it away from the leading edge and then lifting up. The spot welded bracket and its two "ears" do not need to be removed from the base of the stereo. The Faria gauges are just less than three inches in mounting depth and will not interfere with the tray bracket.

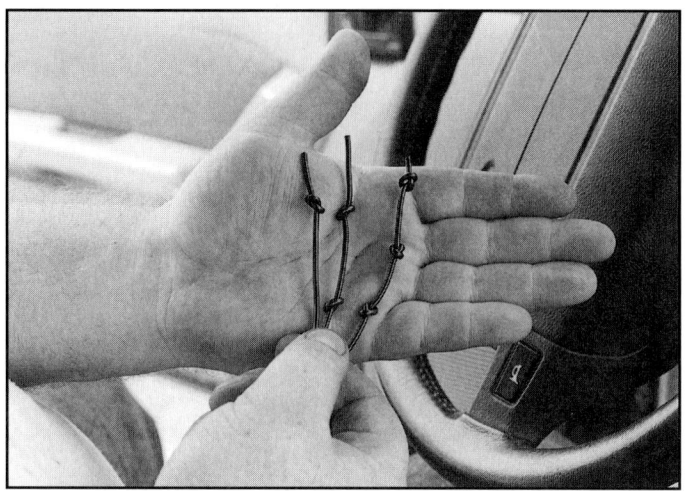

8. You will need three lengths of wiring for the three gauge senders. We made all three lengths the same (since we wanted to have plenty of wire for routing) and coded them to the three gauges by tying knots in them, according to the gauge they belong to, counting from left to right looking at the gauge faces. Crimp an eyelet to one end of each wire and place your knots in the opposite end.

9. Before installing the three sender wires, complete the three gauges' wire connections: power, ground, and lights. Using Scotch-Lok connectors, connect the power lead to the Yellow with black hash or stripe, and connect the gauge lights to the light blue with red stripe, both of which are at the back of the stereo. The ground wire can be mounted to one of the front console mounting screws (arrow), as shown here.

10. Connect the three sender wires to the remaining stud on the back of each gauge and run them to the front of the console at the firewall, keeping them tied away from sharp edges. Plastic wire convolute would work nicely. We made our gauge mounting plate with raised bosses on both the left and right sides to eliminate any side to side movement and thus eliminating the need for mounting screws and making for a cleaner installation. The stereo's lower lip retains the gauge panel too, letting the gauge panel "sit" in its position without any movement. You may wish to use mounting screws or Allen bolts, depending upon your tastes. The stereo will have to be partially installed before the gauge panel can be placed into the console opening. Ensure that the stereo is riding on the upper slide mount for proper support.

AUXILIARY DASH GAUGES

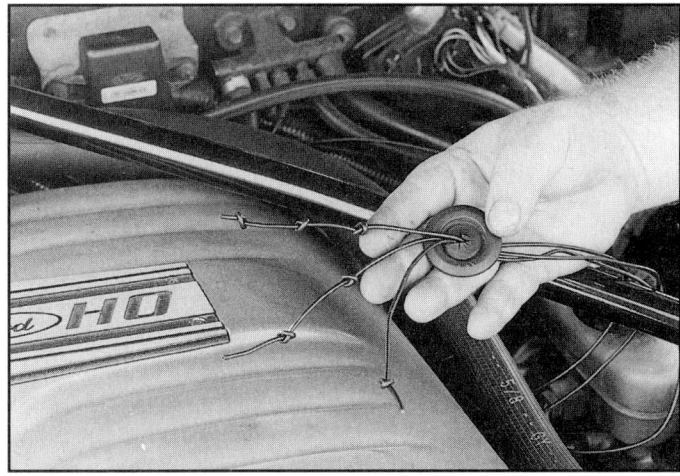

11. Since this 5.0 was equipped with an AOD transmission instead of the ever popular T-5, there is an extra hole in the firewall left for the five speed speedometer cable (on AOD cars the speedometer cable runs through the hole for the clutch cable) just below the BP sensor that we ran our three sender wires through, eliminating the use of a drill and grommets. The factory plug is removed, a small hole cut into it, and the three wires are run through it. With the plug back in the firewall the wires will be out of sight and safely away from any sharp metal. You T-5 owners will have to drill a small hole or run the wires through an existing grommet, such as the factory wire loom by the brake vacuum booster.

12. Instead of trying to cram our cameras all over the engine and take nearly impossible shots, leaving you with a vague idea of where the senders; locations are, we used a demo engine that was awaiting a home in a local dyno facility. The factory oil pressure sending unit is mounted to this extender at the left front of the block. This is the only oil pressure tap that is easy enough for the enthusiast to work with, thus we will have to tee the Faria sender in at this point. A quick call to Faria's 800 tech line told us that the factory Ford senders are very close to theirs and if you don't mind "dead" gauges in the factory dash, you can hook the Faria gauges to the factory senders. But do not try to run the factory and Faria gauges off of the same sender, as this will cut the resistance in half and cause the gauges to read half of what they're really supposed to be reading.

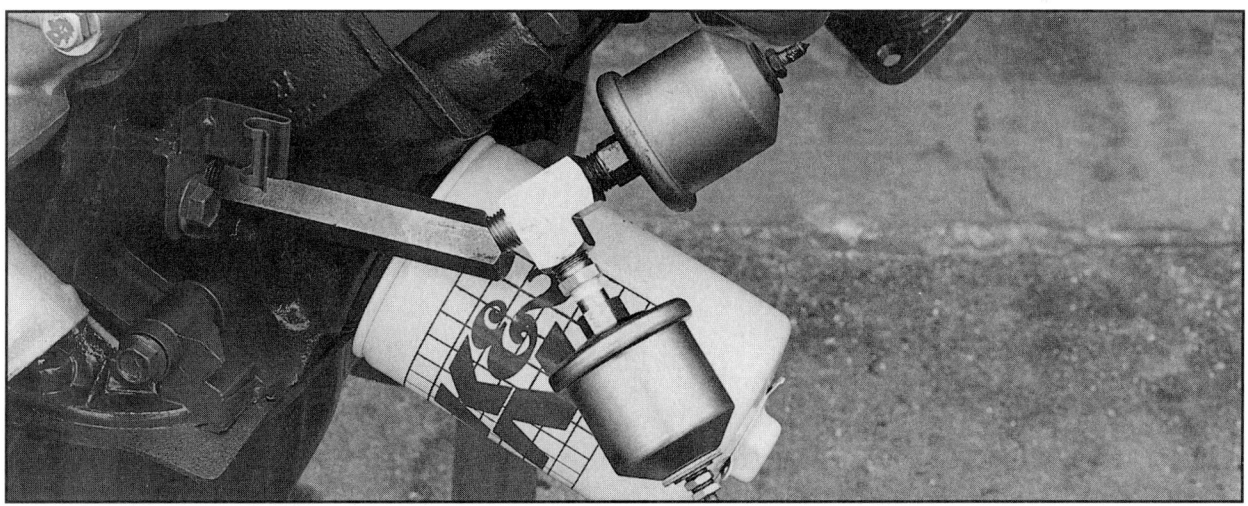

13. A visit to a well equipped auto parts store or a hydraulic fitting company should net you the required adapter, which is a 1/4" NPT "T". You will also need a 1/8" to 1/4" NPT adapter, either from Faria or locally as the Ford oil pressure gauge uses 1/4" and the Faria gauge uses 1/8." Fashioning the adapter and two senders in the manner shown here (with the lower gauge pointing slightly forward towards the power steering pump) will allow for plenty of clearance to mount the original and Faria senders. Use thread tape on all fittings to prevent any leakage. Once completed, cut your appropriate sender wire to the proper length and crimp an eyelet to the end of the sender wire. Complete the sender hookup by placing the eyelet over the sender post and tightening the retaining nut.

AUXILIARY DASH GAUGES

14. The water temperature sender location is at the back of the lower intake. The lower HO intake can accommodate the throttle body and its EGR spacer for either left or right side mounting. To do this the EGR cooler hose will need a port for whichever side the throttle body is on. In this case the throttle body is on the right and the left side hole contains a plug. The plug is a 3/8 square fitting that can be removed with a socket and extension without removing the upper intake. You can easily place your hand between the valve cover and upper intake for installation of the sender. There is no need to drain the cooling system as long as the radiator cap is not removed. It is best to install the sender on a completely cold engine.

15. For the Ford intake it will require another adapter, but this is included with the gauge/sender kit. Again use pipe thread tape on all fittings. Do not over tighten, as brass can easily break. Crimp an eyelet to your sender wire and install it on the sender post as done before. The transmission temperature sender includes fittings for installation in the transmission cooler line, though ours will be in a custom application with a transmission cooler, the instructions are very understandable. This only covers the most common gauges. Others, like a voltmeter, cylinder head temperature, oil temperature, etc. are all very similar to these just installed.

GAUGE PANEL DIMENSIONS

Gauge panel dimensions are 7 1/2" long and 2 1/2" high. A piece of 1/8" thick wood, plastic, Lexan, or aluminum can be used depending upon personal preferences. We used a piece of black Lexan from a sign shop. The Lexan panel was installed with the gloss side in, making the satin finished back almost a perfect match for the matte black dash area. The gauge hole size is 2 1/16" but the standard hole saw sizes are 2" and 2 1/8". Use the 2" size and not the 2 1/8". The 2 1/8" hole saw will make for a loose fit, while the 2" and some minor clearancing with a rat tail file will make for a proper snug fit. The actual gauge placement depends upon how many gauges you wish to install. Some ideas are: three gauges centered, two flush right with a toggle switch on the left, or even one gauge with a toggle switch or other device on each side. The options are many, so decide carefully before cutting. As stated in the text we molded some extra retaining bosses to the back with two-part epoxy for an unobtrusive look to the gauge panel. You may wish to use Allen head bolts or black trim screws with mounting clips depending again on personal tastes.

REMOTE FUEL DOOR RELEASE

For many years the late model Mustang, depending upon trim level and interior options, came standard with a locking fuel filler door and remote electric release. And even if the remote fuel filler door release system didn't come as part of a larger option, it usually was available separately.

In 1990, Ford began making changes to the late model Mustang that would drop this option forever from the Mustang's option list. By reducing product content, or "decontenting" as it is called in the industry, Ford could keep the price on par with the competition and keep CAFE (Corporate Average Fuel Economy) ratings up to prevent Federal penalties from their more gas guzzling lines like the Lincolns. Several items were removed, such as the center console arm rest, remote fuel filler door system, tilt wheel assembly, and other small items. The center console arm rest made it back into production by the end of the year due to complaints by owners, but the remote fuel filler release and tilt wheel options didn't resurface until the new SN-95 Mustang.

Adding the locking mechanism, release solenoid, glovebox switch, and wiring takes nothing more than the removal of a few trim panels, a drill, and some hand tools. We make it easy for you with the instructions below so no wiring books will be needed on your part.

So, if you have a '90-'93 Mustang, or an earlier Mustang that never received the remote fuel filler door release option, and you would like to add the option to keep your 93 Octane Premium yours, the parts are all available right from Ford, or your nearest salvage yard. However, if you do purchase your parts from your Ford dealer, keep in mind that some of these parts listed in this story come two or three per package and some dealers won't break up the packaging these parts come in. In this case, you may be able to find some other Mustang owners in the same situation to purchase the parts together. Otherwise you will have to try and sell the leftover packaged pieces on your own.

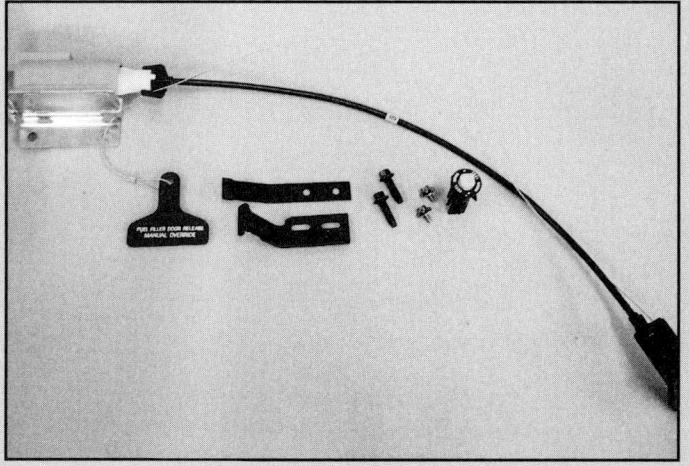

These parts, all available through your Ford dealer, make up the fuel door locking system previously available on the '89 and older Mustang.

Fuel Door Parts List:
Obtain from your Ford dealer:

1	E4ZZ-6128610-A	Solenoid	pkg. of 1
1	E3DZ-5428608-AA	Latch	pkg. of 2
1	E3DZ-54405A24-AA	Latch Spring	pkg. of 4
1	E3AZ-9B242-A	Switch	pkg. of 1
1	N802761-S	Screw Kit	pkg. of 4
1	N605892-S	Screw Kit	pkg. of 6
1	15-20 ft. roll of 14 gauge wire		

REMOTE FUEL DOOR RELEASE

1. Open the glove box and push the two sides inwards to allow the glove box to drop down out of the way. Clear away any loose wires or vacuum lines from the edge of the console and mark your spot to drill a 5/8" hole. This can be accomplished with either a hole saw or a drill bit. An older Mustang with the remote option can be used as a template for your release button.

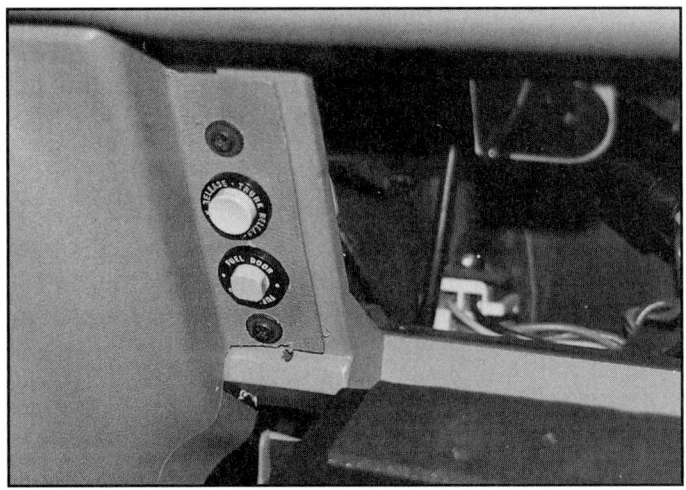

2. Once the hole has been made you will need to make a slight notch at the twelve and six o'clock positions in order to install the switch into the dash.

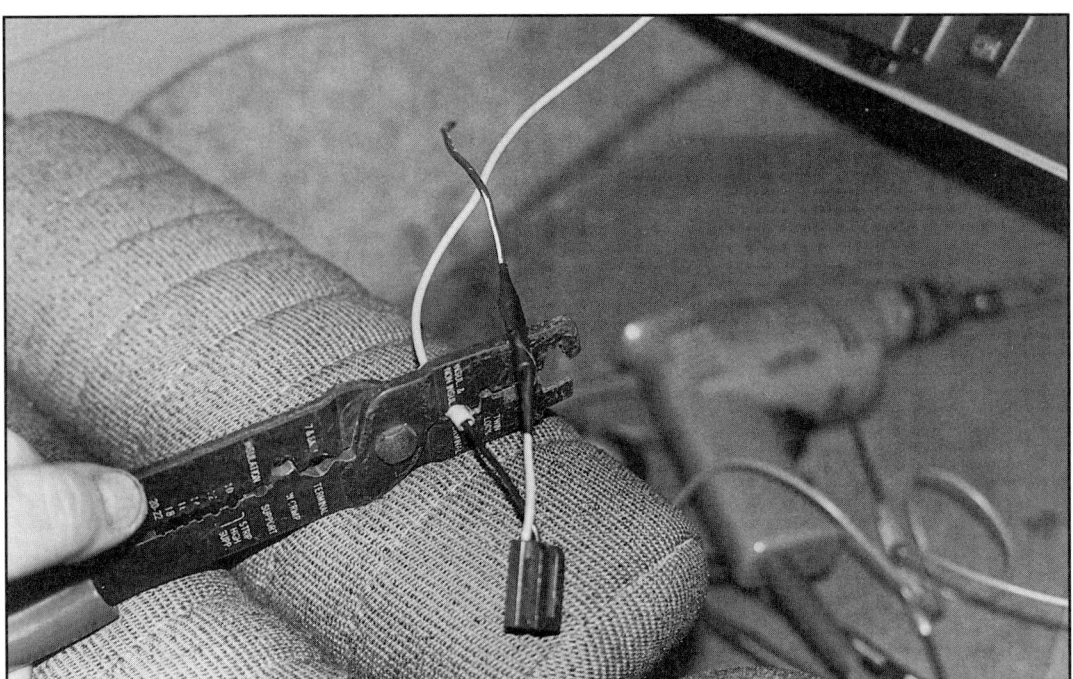

3. You should use the two factory electrical connectors for the switch and for the solenoid. A good place to try for these are a junk yard or a dealership technician. If they have a collection of loose connectors make sure you bring the parts to match them to their connectors. If all else fails you can crimp solderless terminals onto your wires. Here we have begun to run our length of 14 gauge wire (approximately 15-20 feet) and are making the final connections to the connector for the release switch. You can get your power from the glove box light connector (light green with yellow). There is no polarity to the release switch; simply connect one wire to power and the other to the length of wire going to the solenoid.

REMOTE FUEL DOOR RELEASE

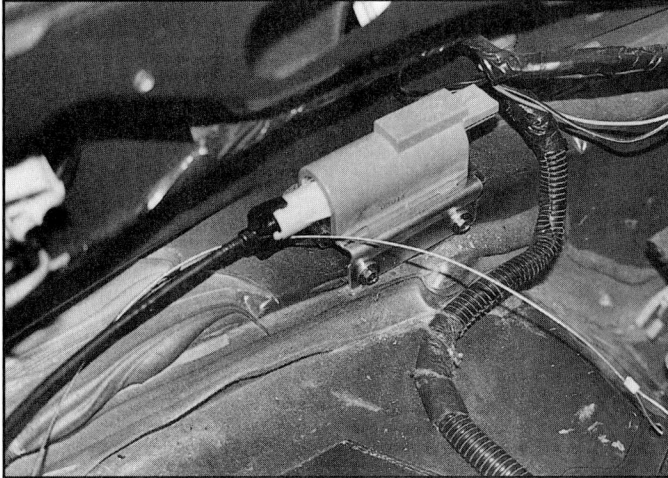

4. Remove the right door sill plate and any other rear seat trim necessary to run the length of 14 gauge wire to the solenoid. The right rear trim cover will have to be removed on hatchbacks for access to the fuel door housing.

5. Mount the solenoid to the lower right taillight panel with two of the four self tapping screws.

6. Measure the holes necessary to attach the release cable, two screw holes and one plunger hole, and drill them out. Make sure that your release cable is centered in the flat mounting pad of the plastic housing. Use two of the six long bolts and attach the cable and plunger to the plastic housing.

REMOTE FUEL DOOR RELEASE

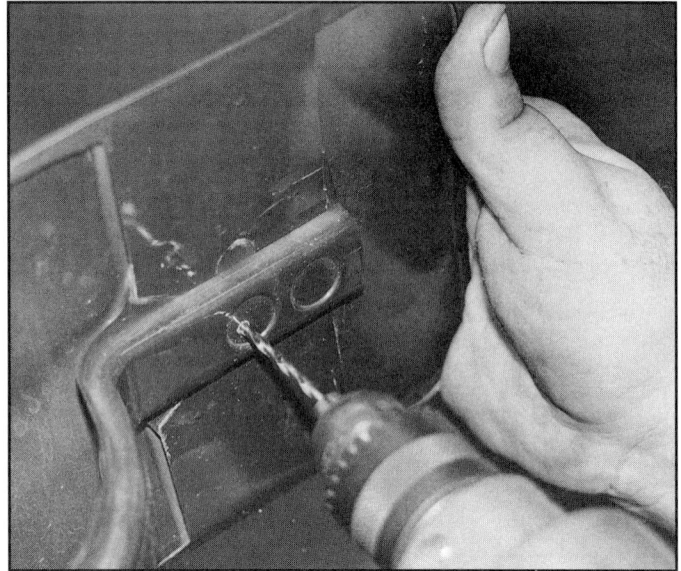

7. This next step must be done carefully. Drill two small holes into the fuel door hinge only! Once the drill bit comes through the hinge pull it back immediately to prevent damage to the outer fuel door.

8. Next, mount the door lock and the spring to the fuel door hinge with the last two self tapping screws. There may be a need for adjustment back and forth to make the door latch shut. Check for proper operation and adjust as necessary.

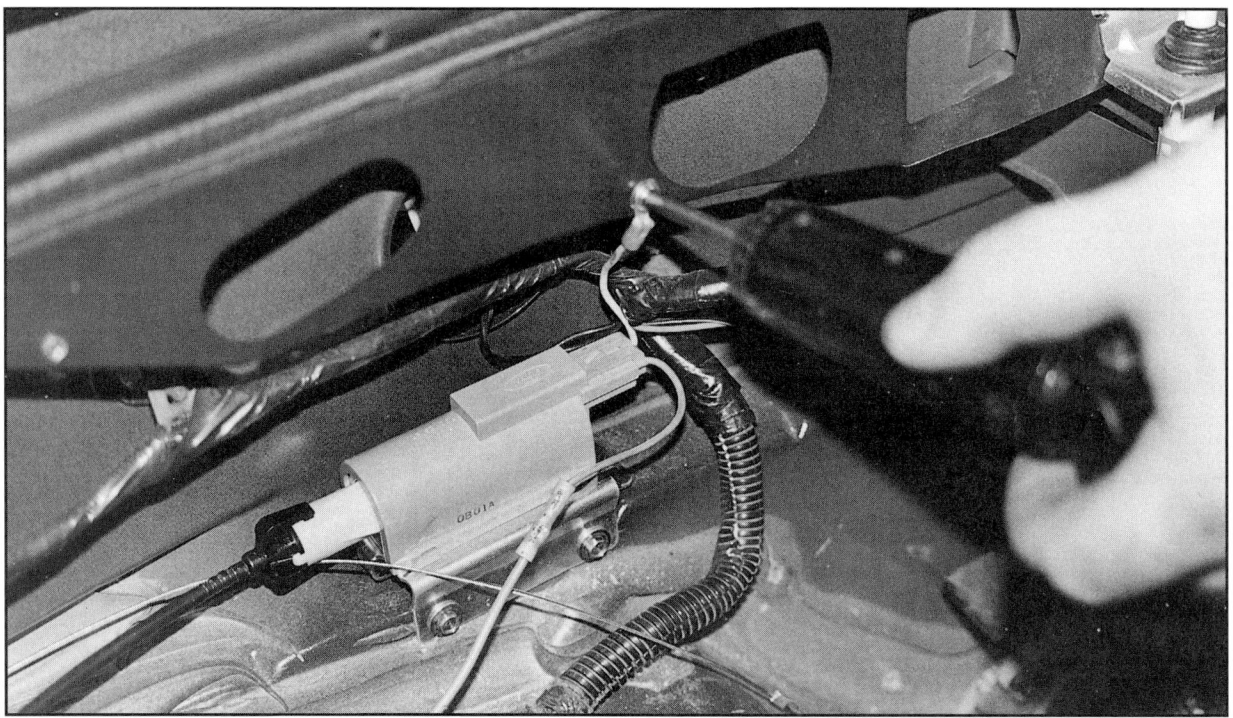

9. Splice the length of 14 gauge wire to the solenoid and the other terminal to ground. The solenoid also has no polarity and can be wired either way. Place the manual release in a location where you can access it and reinstall all interior trim items.

VIII. BODY TECH

INSTALLING FOG LAMPS & REWIRING THE FACTORY HARNESS

SWAPPING TAILLIGHTS

INSTALLING A POWER ANTENNA

COBRA GRILLE INSERT

INSTALLING FOG LAMPS & REWIRING THE FACTORY HARNESS

Light up the night with some aftermarket fog lights for your LX and repair your factory wiring before it causes any major damage.

The '87-'89 Mustang GT has a recurring problem with their headlights, which seem to turn off and on during extended driving with the fog lamp switch also on. Let's describe how the circuit works first, then we can investigate a repair procedure.

The simplest way to explain how this circuit works is to follow the path of current flow from the battery through the headlight switch to the fog light switch. The circuit starts at the starter solenoid "hot" side and travels through a 16 gauge black with orange striped fusible link to the fuse panel where the circuit is divided into several "hot at all times" circuits, one of which is a black with orange striped wire to the main headlight switch, where it passes through a circuit breaker. Only after the headlights are turned on (low beam only) will the circuit continue through a red with yellow striped wire to the multi-function switch (turn signals, flash to pass, etc.). Leaving the multi-function switch is a red with black striped wire that leads to the headlight assemblies. Spliced off of this wire is another red with black wire that leads back to the fuse panel to fuse number 15 (15 amp light blue). From this fuse is a light blue with black wire that leads to the fog lamp switch assembly. Then a tan with orange striped wire leaves the fog lamp switch and goes to the fog lamp assemblies.

Many people have complained of intermittent headlight failure with their fog lamps on. There have even been reported cases of instrument panel fires from overloaded fog/headlight switches. A simple solution to this problem is to rewire the existing small gauge wiring with a heavier gauge, like using better jumper cables.

According to Ford TSB (Technical Service Bulletin) #89-17-11, the culprit is circuit #15. The red with yellow striped wire from the headlight switch to the multi-function switch isn't big enough. You will have to inspect the circuit and replace the defective wiring, and the headlight and multi-function switches if they are damaged. The TSB states model years '87 to '89 GTs, but we decided to play it safe and replace our '90's anyhow while installing new fog lights.

For those LX owners who wish they had fog lights, listen up. If your car is a 5.0, the fog light wiring is also installed in your car. The GT and the 5.0 LX use the same harness. The 2.3 four cylinder is the only one which does not have complete fog light wiring (sorry guys). If you want to install fog lights on your LX, just follow us. After we rewire our '90 LX we will show you LX guys how to install a set of factory "Marchal" fog lights (such as those found on pre-'87 GTs) so you can light up the night like any GT.

The rewiring repair procedure is outlined in steps 1-9; the installation of the Marchal lamps is outlined in steps 10-24.

INSTALLING FOG LAMPS

PARTS LIST

Obtain from your Ford dealer for installation:
(2) E6FZ-15200-A   Fog Light
(2) E6LY-15266-A   Attaching Brackets
(1) E7ZZ-11654-B   GT Headlight Switch

Obtain from your local hardware store:
(4) 3/8"   COARSE X 2.5" long bolts
(4) 3/8"   coarse lock nuts
(4) 1 1/4"   OD washers
(1) 6" PIECE OF 5/8"   OD steel tubing

Obtain from Ford dealer for fog light repair:
(1) E7ZZ-11654-B   GT headlight switch
(1) E8ZZ-13K359-A   multi-function switch
(1) E69Z-14489-B   connector-headlight switch
(1) E6DZ-14489-P   connector-multi-function switch
(2) D8TZ-14474-A   terminal

Obtain from your local auto parts store:
Three to four feet of 12 gauge wire

1. Begin by removing the headlight switch from the dash assembly. Push in on the lock tabs with a pocket screwdriver while pulling the switch toward you.

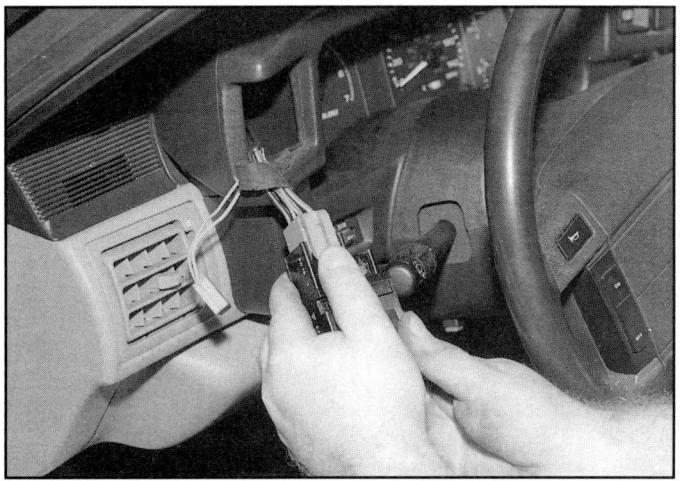

2. Remove the two electrical connectors, the fog lamp by prying with a screwdriver and the headlight by pushing down on the connector and pulling the connector off. NOTE: If the connector is melted to where you can't unplug it from the switch, you will have to break it apart with pliers and install the terminals into the SAME locations on the new connector. It would be wise to write down the wire color and location before you pull any wires out. Remember, if you are replacing the whole connector, swap one wire at a time.

INSTALLING FOG LAMPS

3. Using a small pick, push the tab on the wire connector back (left) and pull the wire out of the connector (right) for the headlight switch.

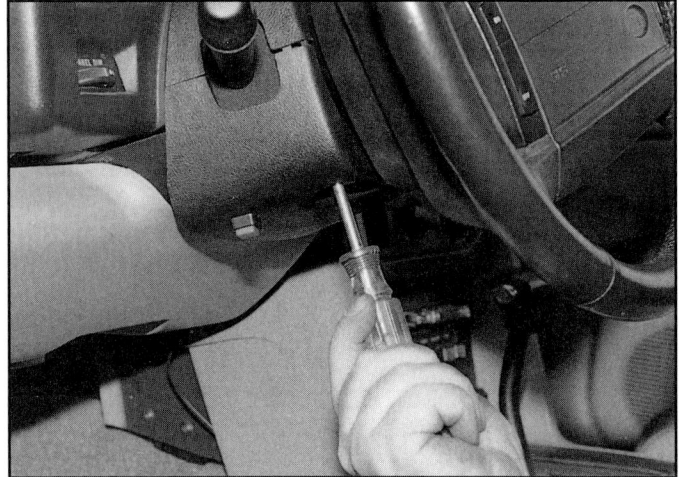 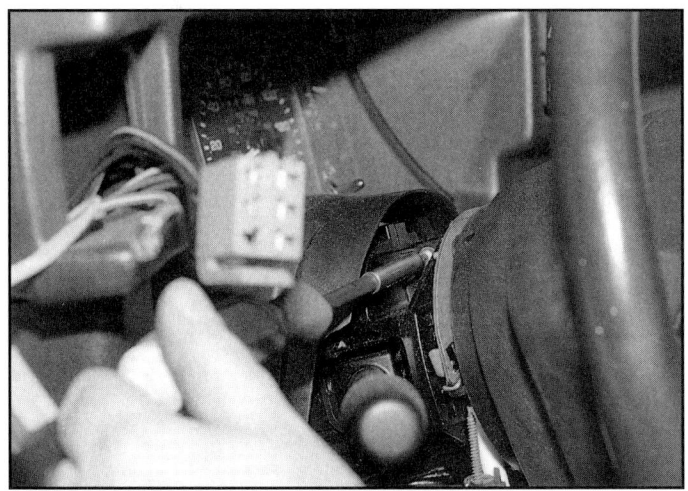

4. Next, remove the knee pad (left) and the steering column shroud to gain access to the multi-function switch and remove the switch (right) from the column.

INSTALLING FOG LAMPS

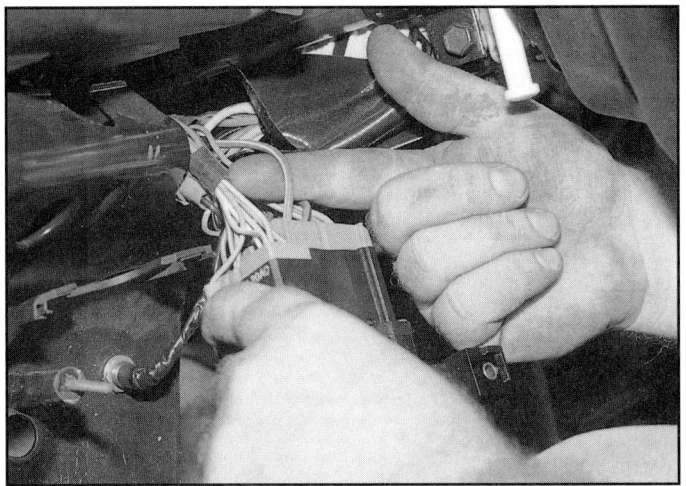

5. Locate the red wire with the yellow stripe in the four wire connector at the multi-function switch.

6. Remove all the connectors from the multi-function switch so you can install the new switch when ready. Don't forget to reinstall the foam ring around the turn signal stalk.

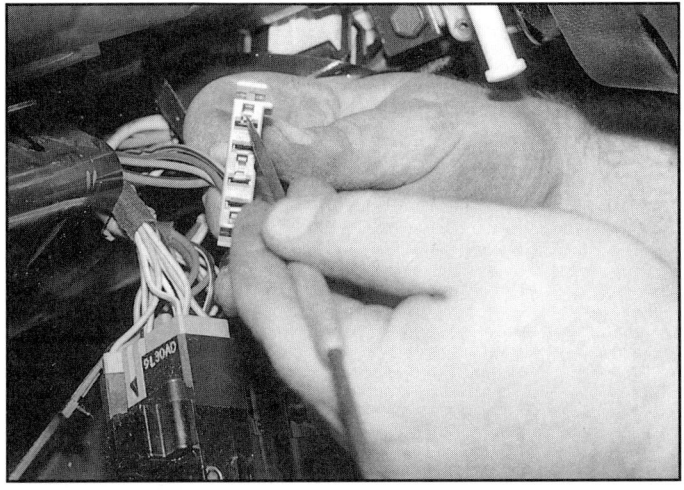

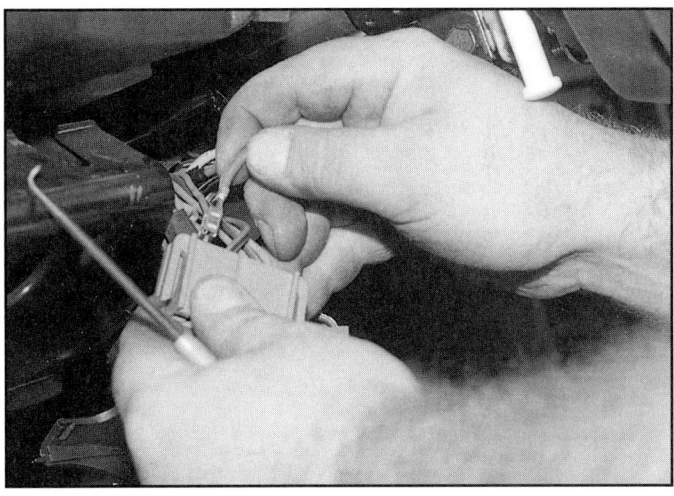

7. Using a small pick, push the tab on the wire connector back (left) and pull the wire out of the connector (right) for the multi-function switch. After you have removed both terminals, cut them off as close to the harness as possible and tape the cut ends.

INSTALLING FOG LAMPS

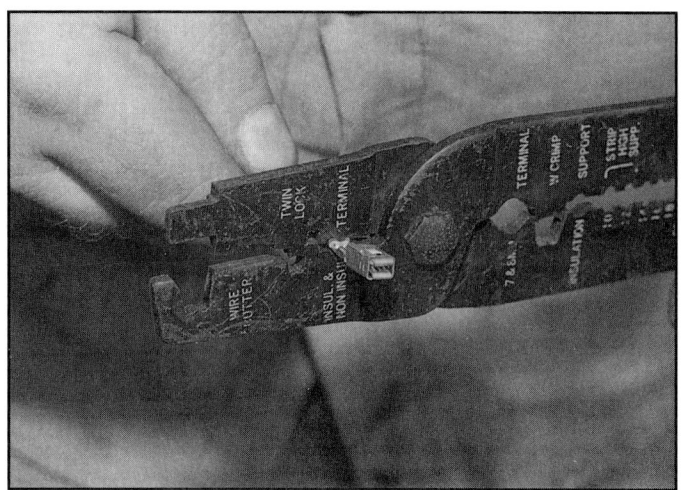

8. Take the new length of 12 gauge wire and crimp the headlight terminal to one end only. Then install it in the connector and run the wire inside the dash (following the old harness) to the multi-function switch. Measure how much wire you will need at the multi-function switch and cut off any access. Crimp the terminal for the multi-function switch onto the end of the new wire and install it in the connector.

9. Install the new headlight and multi-function switches and put the steering column back together. Remember, don't bypass the headlight switch to run the fog lights in the parking light position. Not only is it illegal in most states, but it isn't safe either.

INSTALLATION

Now follow us along on how to install a set of Marchal fog lights on our '90 LX. Remember, if your LX is an '87-'89 you will have to do the previous wiring repair first outlined in steps 1-9.

10. Begin by removing the headlight switch, as outlined above, and install the new GT headlight switch in its place.

INSTALLING FOG LAMPS

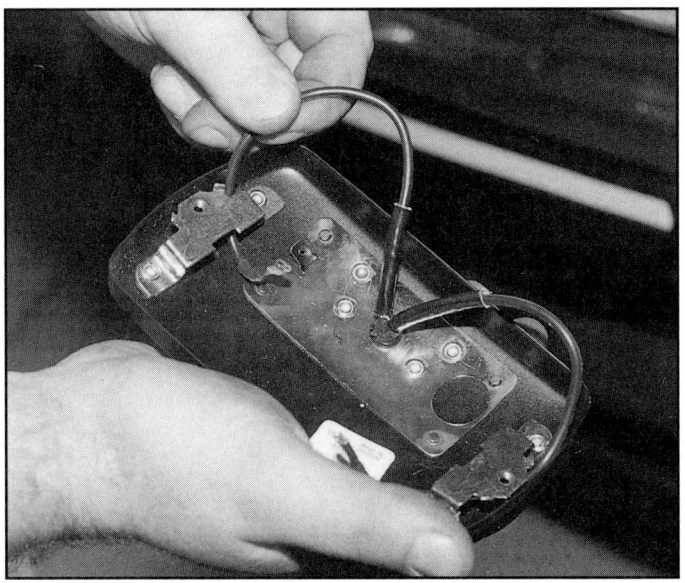 

11. For the fog light wiring to reach the harness without cutting the wiring, we disassembled the lamps (left) and pulled the wiring through the grommet on the back of the lamp to make it longer, and then we reassembled the lamp (right).

12. The brackets and lights come separately, so we decided it would be easier to install the lights on the brackets first, but you can do it either way.

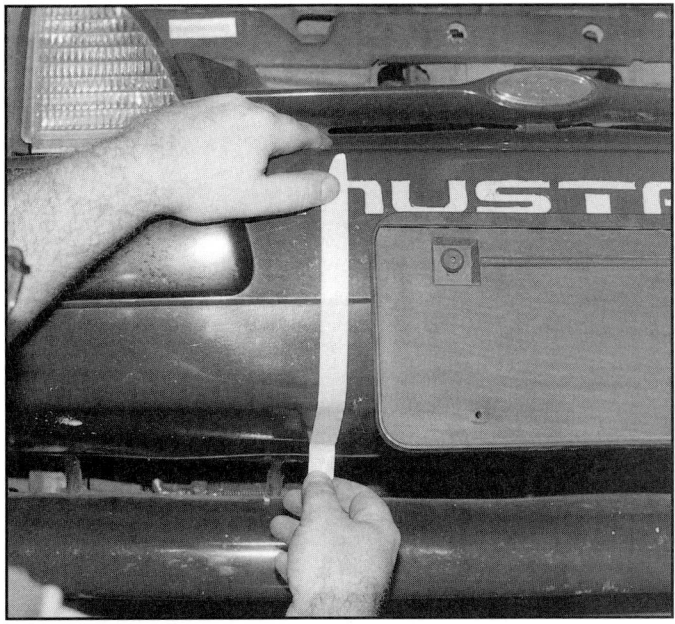

13. Use a length of masking tape as a plumb line from the center of the "M" and "G" in MUSTANG.

INSTALLING FOG LAMPS

14. Place the fog light and bracket against the bumper and mark the inner hole at the tape line and the outer hole accordingly.

15. Drill the two marked holes with a 3/8" drill bit through the fascia and the bumper itself.

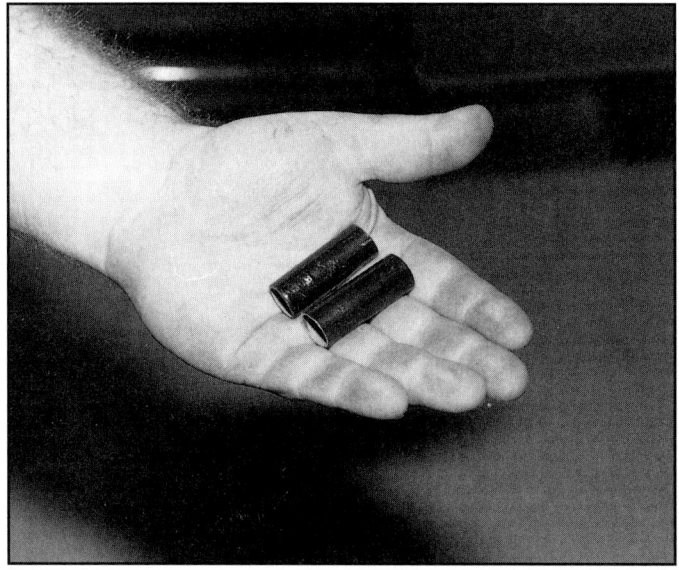

16. Cut four spacers out of the 5/8 tubing. Generally you will need them to be one inch to an inch and a quarter, and the out-board spacers should be approximately a quarter inch taller to keep the lamps level (the fascia curves up at the outer edges). They will serve as a spacer between the bumper and the fascia, as not to deform the fascia when you tighten the bolts down.

INSTALLING FOG LAMPS

17. It might take a little bit of patience to get the spacers to the right length. If you cut them too short, don't panic—just add a washer or two to equalize it. Install the spacers between the fascia and bumper so that the spacers' opening lines up with the holes you have drilled.

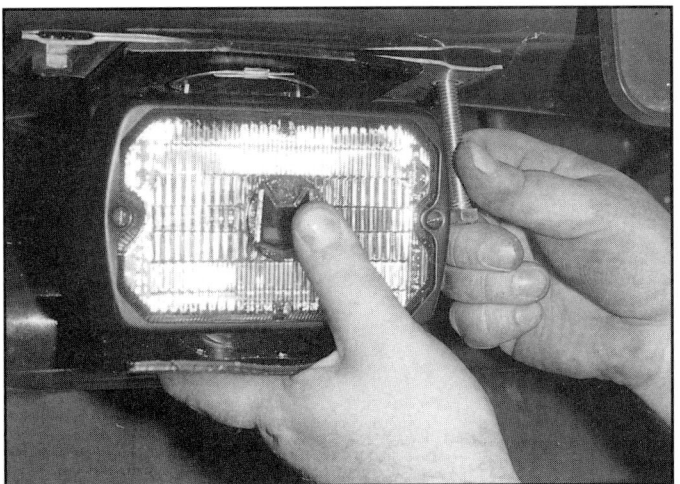

18. While holding the light with one hand, start one of the bolts.

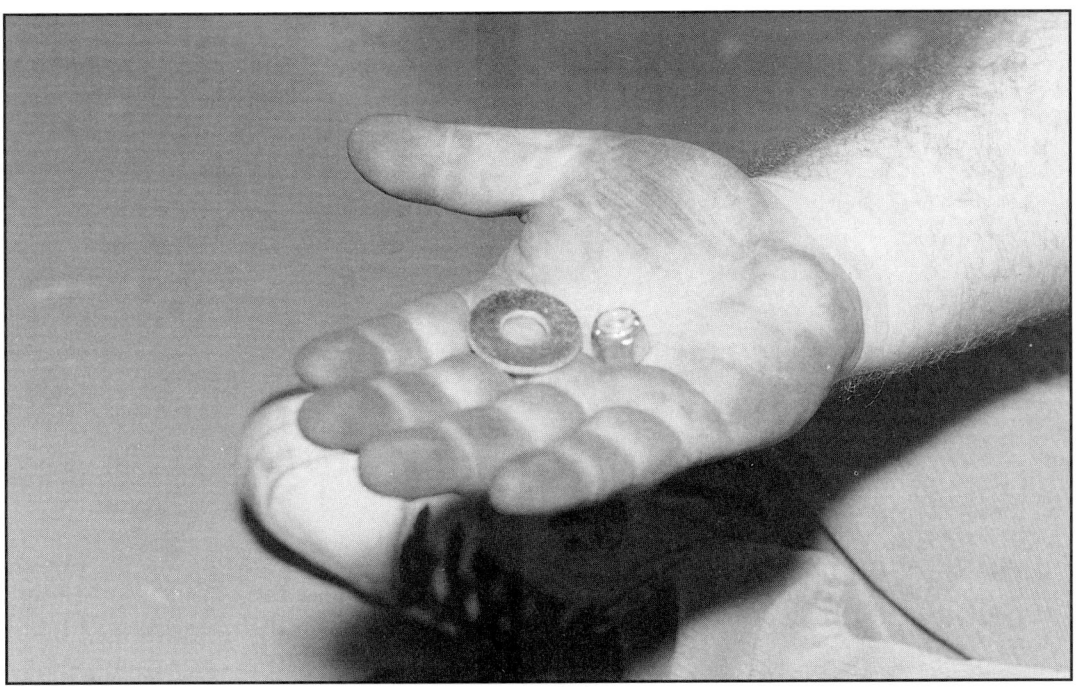

19. Once you catch the first bolt, install the lock nut and washer finger tight.

169

INSTALLING FOG LAMPS

20. Do the same for the other side and then tighten them all evenly.

21. To be able to plug the lights into the factory harness, you will have to pull the fog light harness retainer out of the sheet metal at each front corner to be able to give the harness a couple of extra inches of play. If you are installing non-Ford lights, you can cut the plug off the end and extend the wiring, instead of trying to move the harness around.

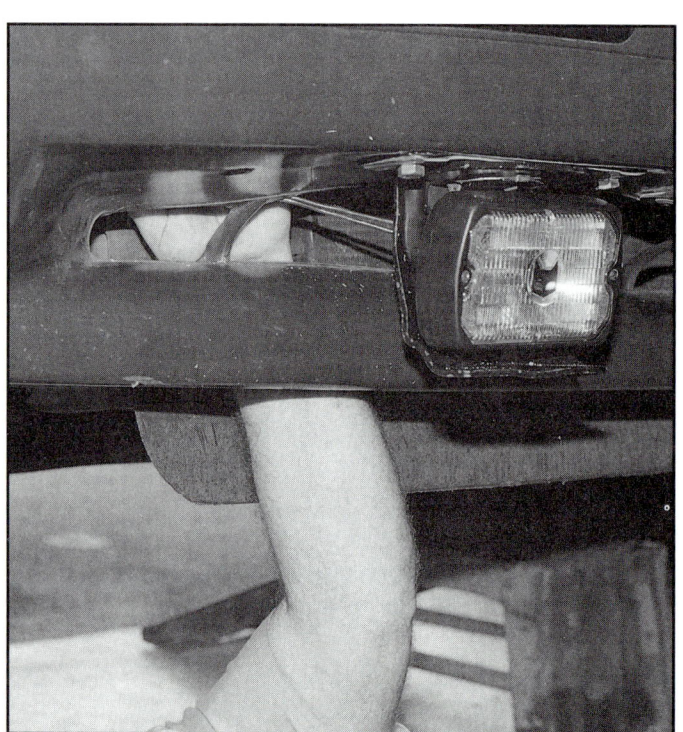

22. Plug the lights into the harness and turn on the headlights (low beam) and the fog light switch and check for proper operation. Then, all you have to do is adjust the lights for proper aim and you're all set.

SWAPPING TAILLIGHTS

It seems like yesterday, but it was almost three years ago that the SN-95 Mustang debuted. I was lucky enough, as a member of a local Mustang club, to see the Mustang up close before the general public at one of the many club showings that Ford put on back in late 1993. The debut was being handled by the Orlando Ford district office and they had Rick Titus, son of Jerry Titus who drove for Carroll Shelby in the late '60s, as lead speaker. Rick, himself a race driver, was going over the attributes of the new Mustang, especially the retro-look of the side scoops, horse in the corral, etc. But when he got to the three horizontal bar taillights, he made a comment to the likes that the taillight engineers were smoking some illegal substance when they designed these "retro" tail lights, either that or they were installed sideways.

Fast forward to late 1995 when the press got hold of some of the '96 Mustangs and pointed to the new '96 taillights and promptly told the Mustang enthusiasts of the world that these taillights, with their vertical three bar element and now looking more "retro" than ever, should have been on the SN-95 Mustang from day one. Strangely enough, the '96 taillight design was supposed to be on the '94 Mustang, but Ford Marketing didn't think people would like them!

Several people have since tested the waters, and yes you can install '96 taillights on your SN-95 Mustang too. These taillights are available through several vendors, including Ford. We went to Classic Design Concepts (810) 624-7997, one of the first companies to try the new taillights for fit, and ordered a set in Rio Red to match our '95 GT. Ordering them this way makes for a true bolt-on, as no painting is required and the wiring harnesses, bulbs, and sockets are included with the lamp kits.

The second part of this section deals with another aesthetic change. Many owners of '87-'93 GTs would give anything to get rid of that "cheese grater look" on the taillamps. The switch to the smoother, rounded LX look is easy enough to accomplish, as outlined on pages 174 and 175.

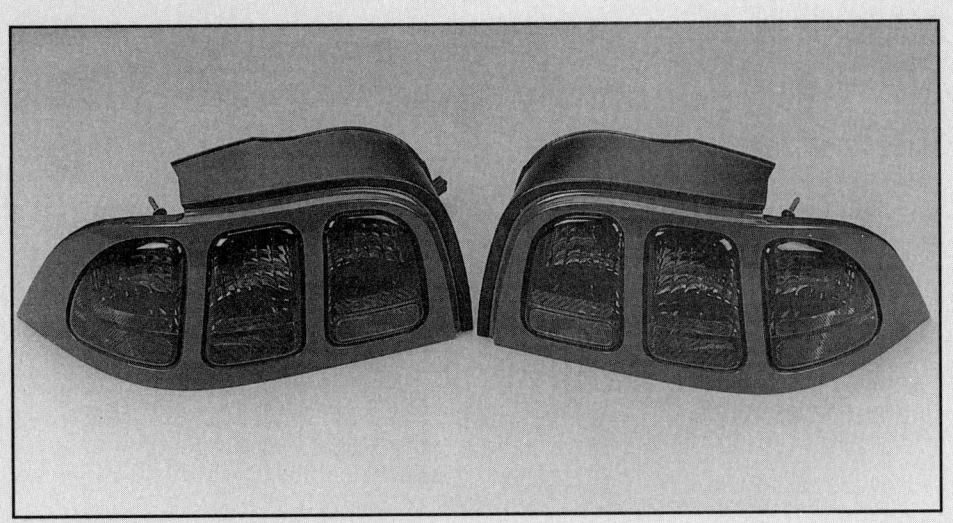

The '96 taillamps come completely assembled with wiring and bulbs; simply bolt them in place on your '94-'95 and plug the wiring in.

SWAPPING TAILLIGHTS

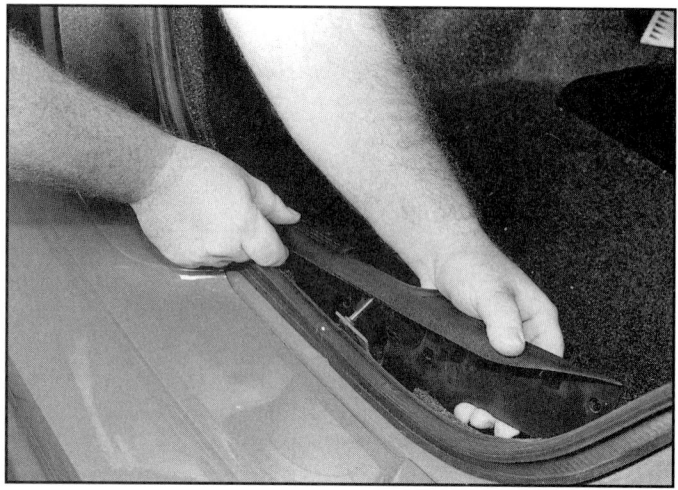

1. The trunk latch guard will need to be removed first to access the trim carpeting. Remove the four retaining plugs with a panel removal tool or pliers and set the latch guard aside.

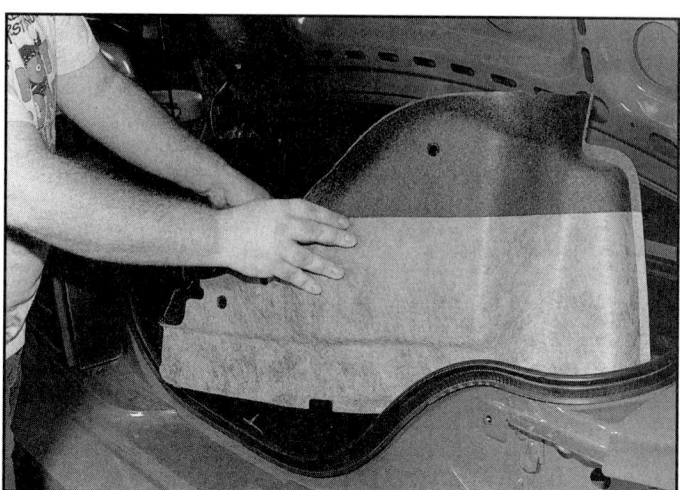

2. There are two push pins and two cargo net hooks to be removed before the rear trim carpet panel can be removed. Careful bending of the panel will get it out of the trunk in one piece.

3. Remove the four metric retaining nuts (per side) and keep them for the '96 lights.

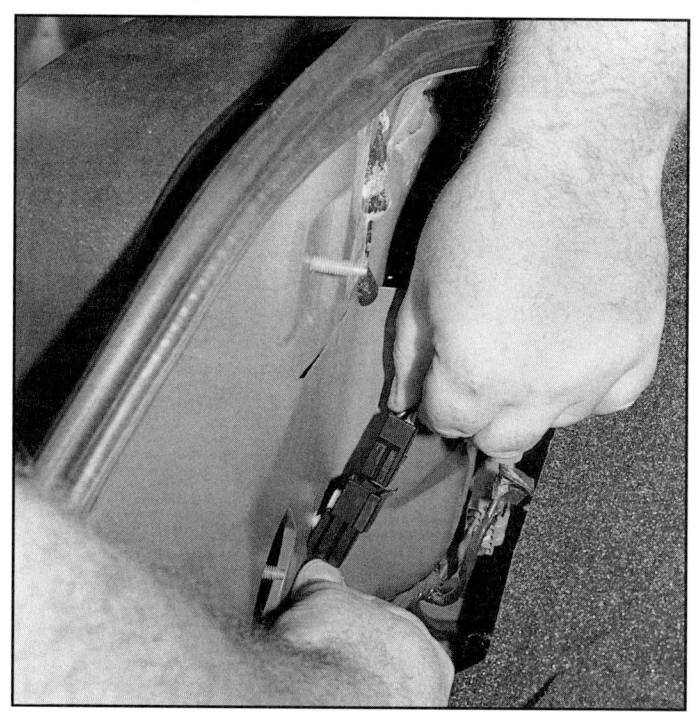

4. Pry back the locking clip of the taillamp electrical connector with your thumb and pull apart the two halves of the connector.

SWAPPING TAILLIGHTS

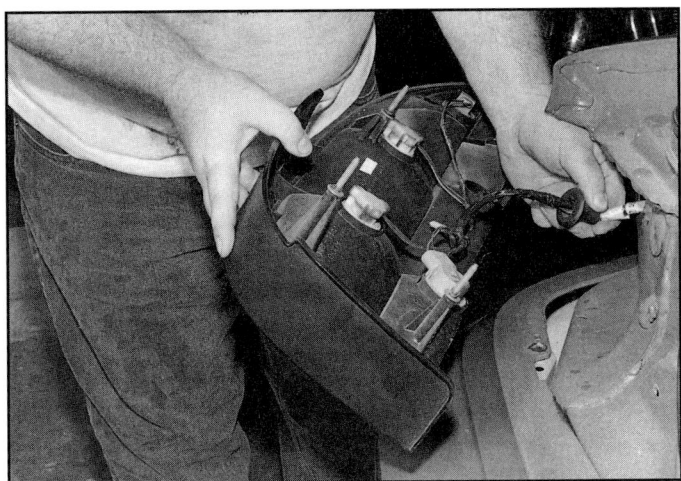

5. Carefully pull the lamp away from the body work until the wiring harness grommet is visible. While holding the lamp in one hand, push the grommet through with your other hand to free the wiring harness from the taillight panel. Set the tail light assembly aside since you will not need any parts from it.

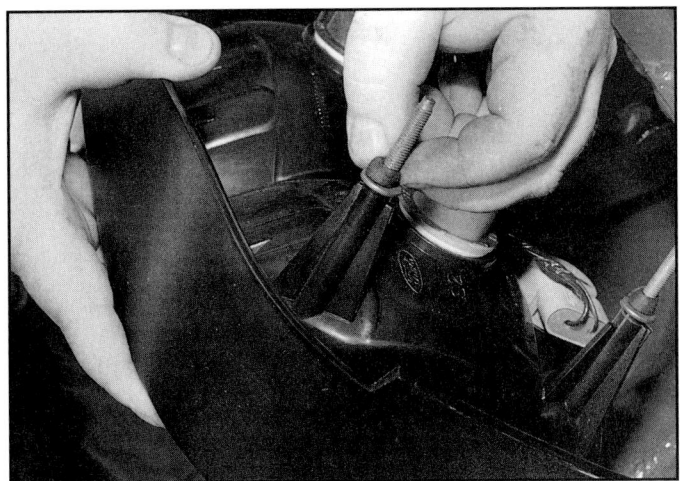

6. Before installing the new wiring grommet, place a small ring of butyl or rope sealer around the four mounting studs of the taillight housing, this will prevent water intrusion into the trunk area. While holding the lamps in place, evenly tighten the four retaining nuts, connect the wiring plug, and reinstall the trunk trim to complete the job.

7. The stock taillight lens assembly is shown here to the left, with the new '96 taillight lens assembly to the right. The '96 lens is unmistakable compared to the older lens. Also notice the jewel-like quality of the lens on the '96. The new lamps use a composite reflector behind the bulb to direct the light in different directions instead of the lens itself doing the job. This gives a much more expensive look to the Mustang and makes for brighter and safer lamps.

SWAPPING TAILLIGHTS

Swapping '87-'93 GT lenses with LX Lenses

For '87-'93 owners of GT Mustangs, swapping to the smoother looking LX taillights is a simple driveway job as well. Watch how we transform this GT at Dugan Racing from the "cheese grater" look to nice LX smooth lenses.

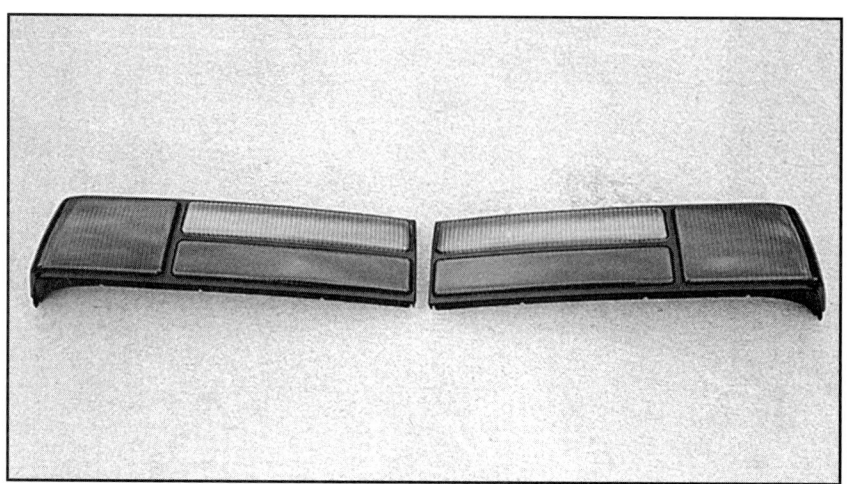

The LX lenses can be purchased separately. There is no need to remove the whole light housing to complete the job.

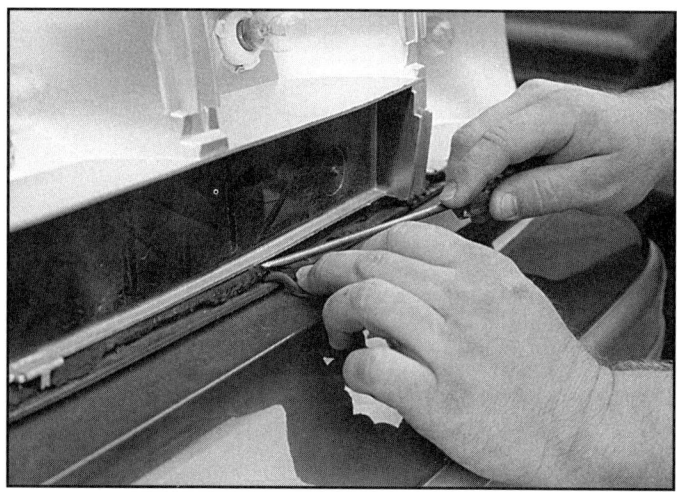

1. Using a flat blade screwdriver, gently pry the lenses away from the reflector housing. Using a narrow tip flat blade screwdriver, scrape away all the old butyl sealer from the reflector housing, being careful not to knock the spring clips out.

2. After removing all the old butyl, clean the area and apply fresh sealer from the supplied roll.

SWAPPING TAILLIGHTS

3. Line the lip of the taillight lens up with the strip of sealer and apply firm, even pressure to the lens to secure it. Do not use a rubber mallet or other tool to seat the lens, you will scuff the plastic or crack it.

4. The GT takes on a whole different look (left) compared to the original lenses found on the GT (right).

INSTALLING A POWER ANTENNA

The late-model Mustang has never come from the factory with a power antenna. Though many other Ford products enjoy the security and benefits of a power antenna, the Mustang has been blessed with a fixed one-piece antenna mast. This one-piece mast screws to the antenna base mounted on the fender, allowing for easy removal by some vandal, or worse yet, to be bent, pretzel-like, into a mass of twisted metal.

Probably the worst thing about a fixed mast antenna is how much the darn thing gets in the way. Want to install a car cover? Now you have to cut a hole in the car cover for the antenna to pass through. In a hurry and wish to run your late-model through an automatic car wash? Not if you want to keep your antenna from being turned into a pretzel.

What can a late-model Mustang owner do to remedy these problems? A power antenna is the only answer. Crutchfield Corporation, well known for their easy to install car stereo kits, also carries a full line of power antennas and mounting kits for many cars, including late model Mustangs. To keep the blacked-out factory look we installed a fully automatic power antenna with a three piece mast that is anodized black for a factory looking installation.

With a fully automatic power antenna the rising and lowering of the antenna mast is dictated by a signal from the stereo head unit. In the case of most all aftermarket head units built today in the "mid-priced" range they have a signal lead strictly for use by the power antenna. In the case of a factory stereo you can use a small 12 volt relay from Radio Shack or other electrical parts vendor to utilize the signal lead to the Premium Sound amplifier as your antenna power lead. If you don't have a factory Premium Sound stereo system, but still want the convenience of a power antenna, you can control the raising and lowering of the antenna mast via a small rocker switch mounted on your console or under the stereo.

For the automatic route, go to Radio Shack and get 1 Cat. # 275-233 12VDC Reed Relay. You will need to solder about four to six inches of 16 gauge wire to the four terminals of the relay. The Y/BK, O/LB, and Red can be found at the inboard wiring connector on the back of the radio. Scotch lok the soldered wires to these, as indicated in the drawing, then butt splice the fourth wire to the green wire coming from the antenna relay. With the antenna wired in this manner it will extend or retract with the radio on/off signal to the amp. It will also retract when the key is shut off.

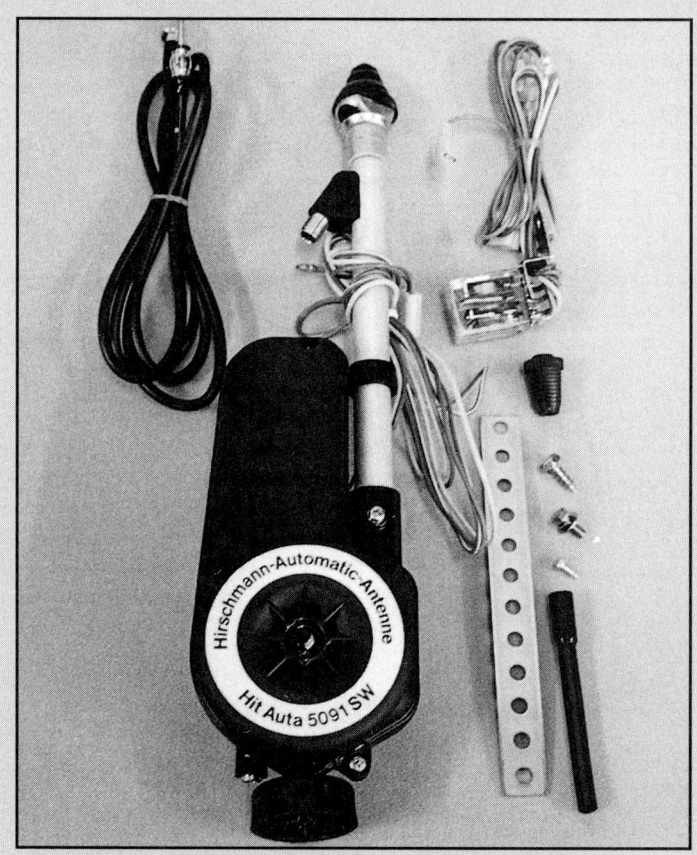

Here is our antenna kit from Crutchfield. It is manufactured by Hirschmann, a supplier for original equipment antennas for Ford and other major manufacturers.

INSTALLING A POWER ANTENNA

1. You will need plenty of front fender access. Jack up the right front of your Mustang and support it with a jack stand. It will be necessary to remove the tire to access the fender liner. There are three screws along the perimeter of the fender lip, along with two screws directly beneath the rear edge of the front fender and one under the front edge. Once these screws have been removed it is merely a job of removing the various plastic retainers and pulling the fender liner out of the fender area.

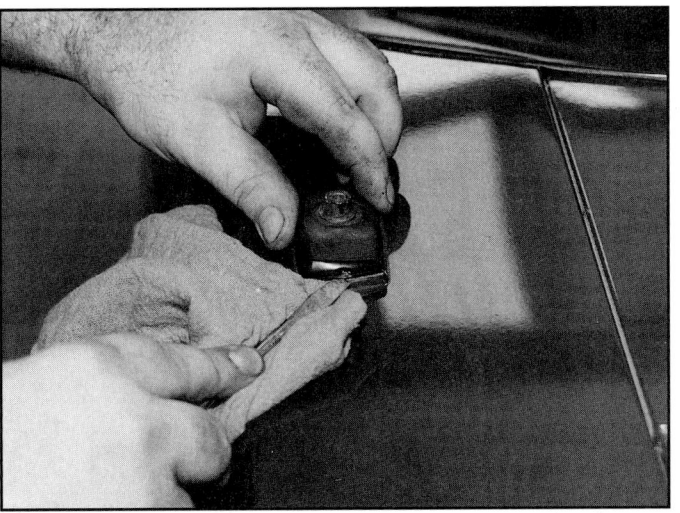

2. Using a 3/8" boxed end or flare nut wrench, remove the old fixed mast antenna. A small rag under your screwdriver will protect your paint while prying the antenna base cover off of the antenna base. With the base removed, you now have access to the four small Phillips retaining screws. Remove these four screws but don't remove the antenna base at this time.

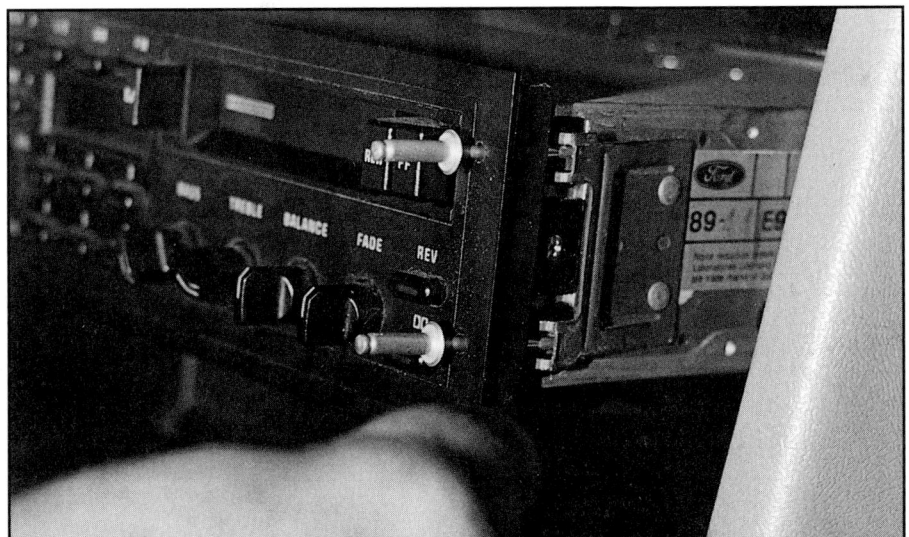

3. Ford uses a locking device for their DIN chassis stereos. A dealer technician would use a special service tool. This tool costs over $20 and does the same job as four 1/8" inch nails. We used four 1/8" inch rivets because their head made for a more comfortable pressure point for the fingers. Insert the four nails or rivets into the holes with the tips at a slight outward angle. Once inserted, push the four nail heads away from each other (outward) and pull the stereo out of the console. Patience is a real virtue while doing this maneuver and it may take an extra hand from a helpful neighbor.

INSTALLING A POWER ANTENNA

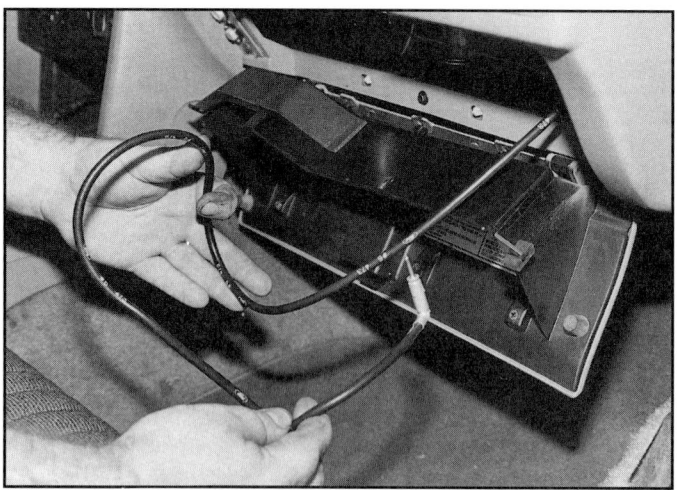

4. Remove the right front kick panel to gain access to the inner fender and cowl area. Disconnect the antenna lead from the back of your stereo and route it back through the glove box (push the sides in gently to drop the door open) in preparation for removal from the car.

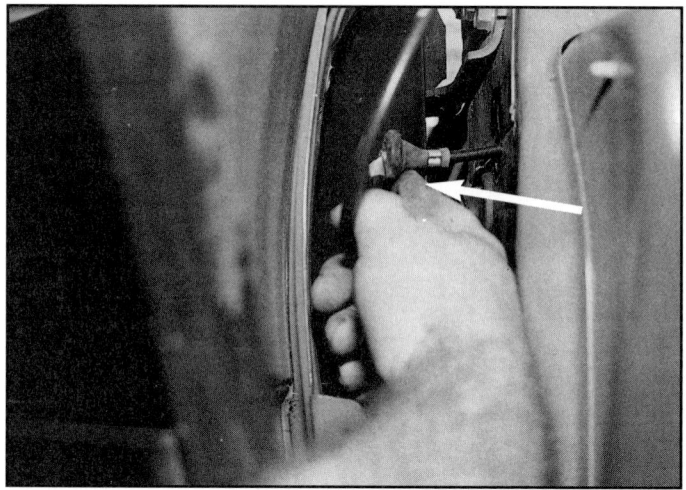

5. Reach up into the back of the fender and pull the rubber sealing grommet (arrow) out of the inner fender. Continue pulling the antenna lead out until it exits the car through the hole. Now you can remove the antenna base from the fender. Be careful and don't scratch your paint with the end of the antenna lead.

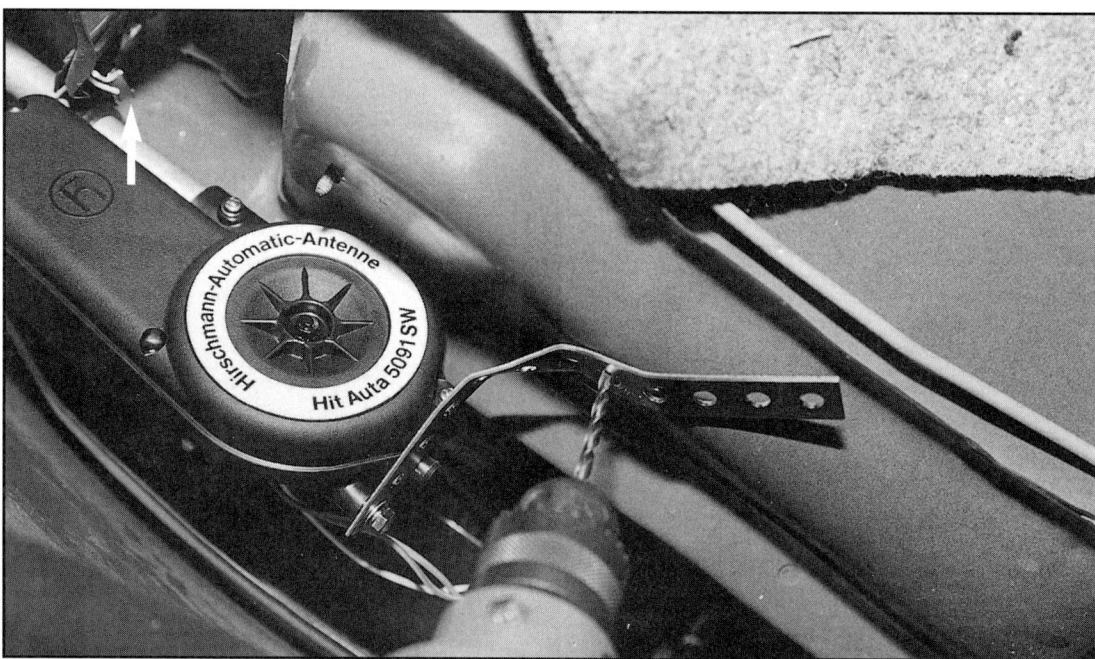

6. Follow the installation instructions for an easy assembly. The arrow is pointing to where we had to slightly clearance the cowl weld for the mast shaft to pass by. We ordered the Ford antenna head kit to complement this antenna and aid in installation.

INSTALLING A POWER ANTENNA

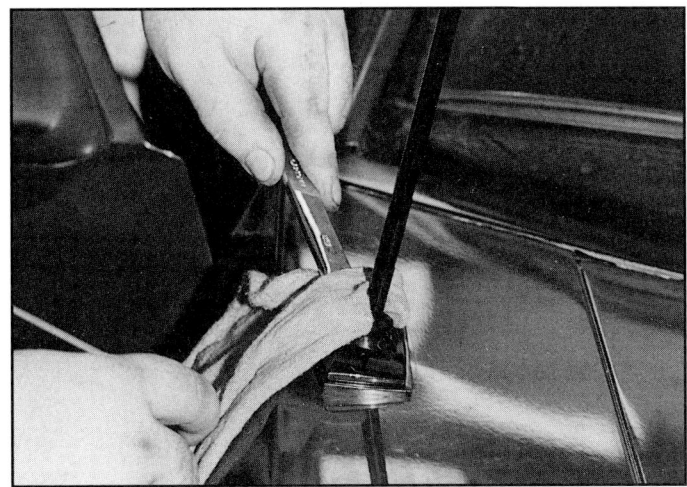

7. Once you have the antenna in position, place the head kit onto the fender and tighten the retaining nut with a cloth covered wrench.

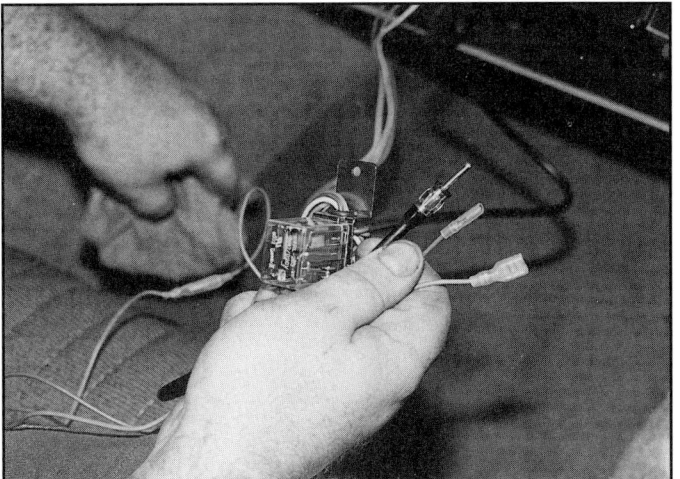

8. Run the new wires and antenna lead through the original hole and into the glove box area. These are the four connections you will need to make: Relay body to chassis ground, antenna lead to stereo, green to stereo or key power, and yellow to battery power (such as green with yellow stripe for the interior lights at the right kick panel).

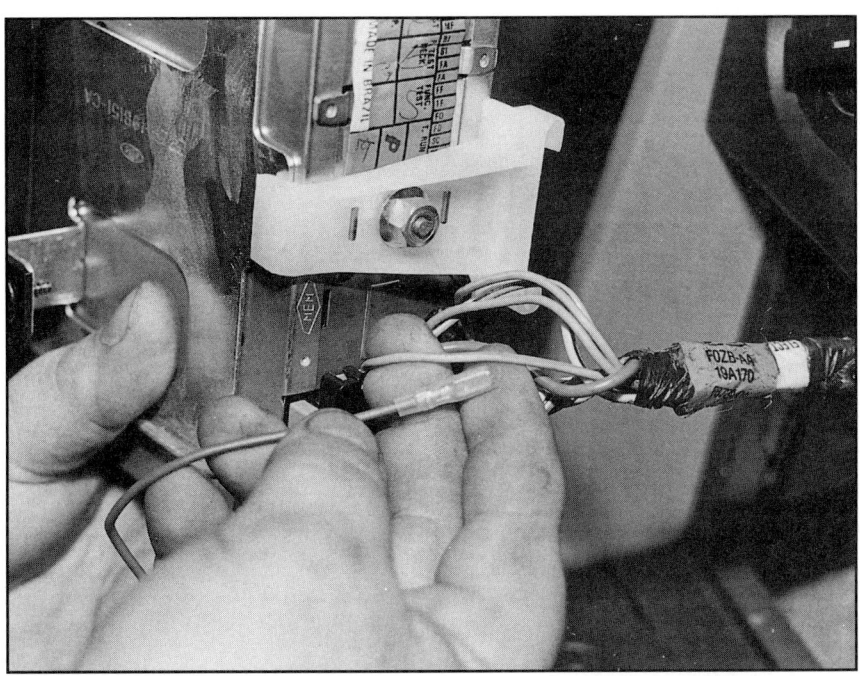

9. This is the orange with light blue wire at the stereo connector. The green wire will have to be spliced into this wire through a relay (see page 176). This wire is only available on premium sound optioned cars. If your car has a base stereo then the green wire will have to be spliced to a key power signal.

COBRA GRILLE INSERT

Most body modifications, such as rear wings, ground effects, and fiberglass hoods, are beyond the scope of a "weekend" project, due to the simple fact painting and body work is required before installation. We, however, feel the Cobra grille insert, found from the factory on the '93 Cobra Mustang, is one small item we could get away with. Most vendors that sell the aftermarket version offer pre-painting as an option, and if you own a teal, black, or red Mustang, you can buy a pre-painted one right from Ford. Even if you buy an unpainted grille insert you can have your local body shop pre-paint it for you in plenty of time for your planned weekend of installation.

No body work is required to your Mustang except for cutting the opening where the Cobra grille insert will reside, and basic handtools are all that are needed for the installation, though an electric or pneumatic cutting tool will aid in the cutting of the front bumper cover. Installation can be accomplished with the front bumper fascia on your Mustang, but a little work and about fifteen minutes are all that it takes to get the front fascia off your Mustang and allow for a hassle free installation.

Most people are familiar with the running horse logo from the vintage Mustang grille. The famous chrome pony became a hood emblem in '79 and finally disappeared altogether in '83. Why Ford decided to use the running horse again after all these years on the '93 Cobra is anybody's guess, but we are glad they did. It looks absolutely great and right at home, the only question is, "How do I make my car's grille look like that?" Short of buying a Cobra Mustang itself, there is one other way. By purchasing an aftermarket or Ford Cobra grille insert you can install this nose-piece into any LX or GT front fascia. The OEM nose-piece is not part of the front fascia, but an add-on insert that you can install in your own driveway. A great advantage to the Cobra grille insert is that it will aid in cooling by creating another grille opening on GT cars. You can also purchase different emblems for your grille insert, such as the Cobra snake emblem, and even a Ford Blue oval. Follow along as we transform an '88 GT convertible with a Cobra nose-piece.

The Cobra Mustang grille insert, this one from Dugan Racing, will add style and distinction to your '87 and newer late model Mustang.

COBRA GRILLE INSERT

1. Our project vehicle, an '88 GT convertible, had lost a fight with the rear of a Ford Bronco, but the Bronco had the unfair advantage of a protruding trailer hitch. Unfortunately our Project GT will need a new nose cover, which Dugan Racing also provided. The steps for installation of the Cobra insert don't require the removal of the nose, and the LX version is even easier to install.

2. Removal of the front fascia is actually easier than you might think. All wiring connections must first be disconnected. This includes the headlights, parking lights, and turn signal lights. On GT models the fog lights do not need to be disconnected as they are not part of the front fascia.

3. There are nine fasteners per side of the Mustang that retain the front fascia and headlight panel. Start with the L shaped bracket just behind the turn signal, then proceed to the lower mounting stud just behind and below the headlight assembly, and finally the five bolts and two nuts that retain the fascia to the fender and fender extension and/or ground effects. Our GT had four self tapping screws, located under the grille opening, that secured the fascia to the fog light bar. The LX nose may have slightly different mounting locations in this area.

COBRA GRILLE INSERT

4. With all the wiring connections disconnected and the 18-plus fasteners removed, gently pull the fascia forward to remove the studs from the fenders. On GT models, beware that the lower ground effects section may need some help over the fog light bar, otherwise it may get snagged on the light bar.

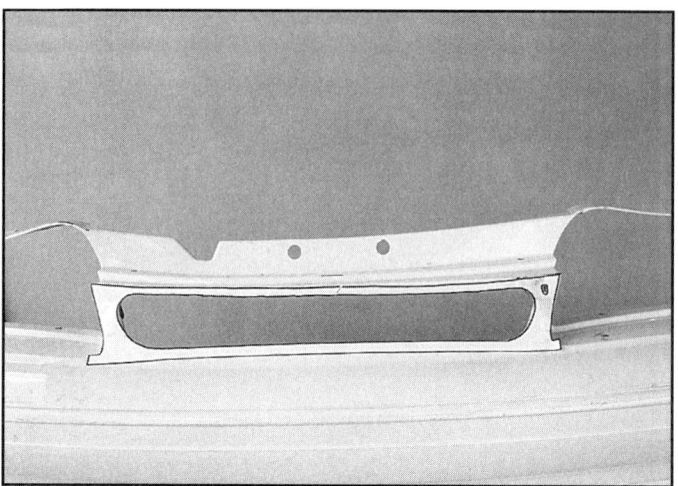

5. Our new fascia has yet to have the headlight panel installed, but yours can be cut for the Cobra insert without removing the headlight reinforcement panel. Line up the provided template with the nose area and then proceed to cut the opening into the fascia with a jig saw or portable saws-all. Once our opening had been completed, the new fascia and the Cobra insert were sent to a body shop for priming and "cutting" of all hidden areas with paint. Again, you can get the Cobra grille pre-painted from some vendors, or you can have the insert painted before your weekend arrives.

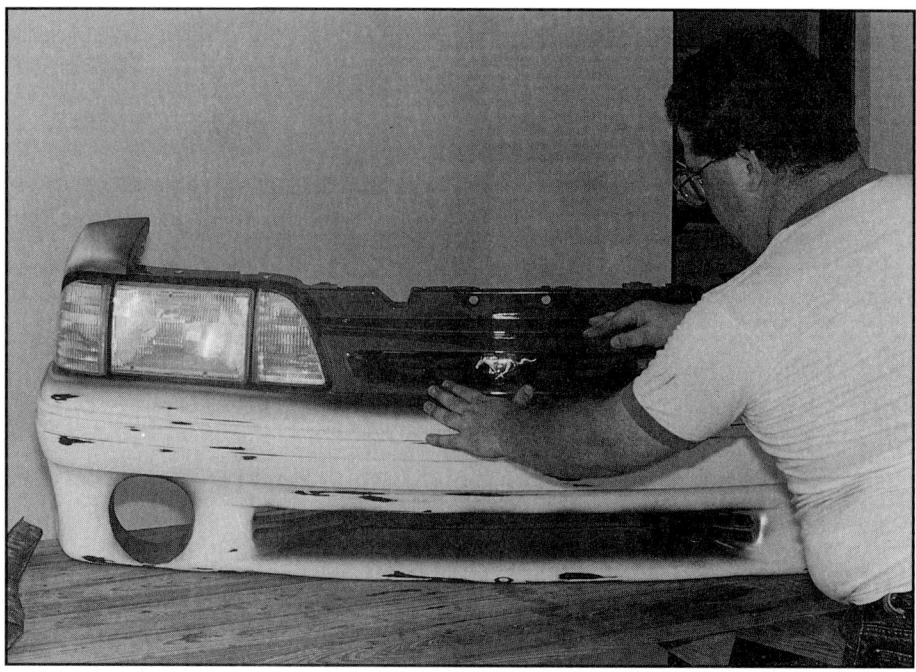

6. Once our fascia had enough time to dry, we reinstalled our headlight panel and all the lights. This operation would not be needed unless you are installing a new fascia, such as we did. Place the Cobra insert into the opening in the grille and center it as needed.

COBRA GRILLE INSERT

7. To install the two attaching screws we used a ratcheting boxed-end wrench with a Phillips head bit inserted into it as sort of a right angle screw driver.

8. A right angle screwdriver, or a small amount of clearancing of the headlight panel, will be needed to allow tightening of the two attaching screws. Double check the insert to ensure that it is still centered as you tighten the two screws.

9. Reinstall the fascia to the vehicle loosely and enlist the help of a friend to align the fascia along the fender and fender extensions as you tighten the bolts and nuts. Once these have been tightened and the fascia has been properly aligned, the remaining bolts and nuts can be installed. Do not forget the small self tapping screws along the fascia's lower ledge or the bottom will flap in the wind at speed. Reconnect all light wiring and stand back and take in your work.

ABOUT THE AUTHOR

Mark Houlahan has been an automotive journalist for over five years as the technical editor for Dobbs Publishing Group's *Mustang Monthly* magazine, based in Lakeland, Florida. He has been a Mustang enthusiast for over 10 years. Mark's love for all things Ford, and especially Mustangs, can be credited to when his parents bought him a '66 Mustang hardtop as his first car. Later, Mark purchased a '90 LX 5.0 from the local Lincoln/Mercury dealer he worked for in South Florida, and he was bitten by the late-model 5.0 performance bug. His love for Mustangs and for writing technical articles led him to *Mustang Monthly*, where he has been since. Mark lives in Central Florida with his wife, Dawn, and their two children, Kyle and Shelby.

OTHER BOOKS OF INTEREST

1,001 High Performance Tech Tips by Wayne Scraba	1-55788-199-5/$16.95
Auto Math Handbook by John Lawlor	1-55788-020-4/$16.95
Automotive Electrical Handbook by Jim Horner	0-89586-238-7/$16.95
Automotive Paint Handbook by Jim Pfanstiehl	1-55788-034-4/$16.95
Brake Handbook by Fred Puhn	0-89586-232-8/$16.95
Camaro Performance Handbook by David Shelby	1-55788-057-3/$16.95
Camaro Restoration Handbook by Tom Currao and Ron Sessions	0-89586-375-8/$16.95
Chevrolet Power edited by Rich Voegelin	1-55788-087-5/$19.95
Classic Car Restorer's Handbook by Jim Richardson	1-55788-194-4/$16.95
Holley Carburetors, Manifolds and Fuel Injection (Revised Edition) by Bill Fisher and Mike Urich	1-55788-052-2/$17.00
How to Make Your Car Handle by Fred Puhn	0-912-65646-8/$16.95
Metal Fabricator's Handbook by Ron Fournier	0-89586-870-9/$16.95
Mustang Performance Handbook by William R. Mathis	1-55788-193-6/$16.95
Mustang Performance Handbook 2 by William R. Mathis	1-55788-202-9/$16.95
Mustang Restoration Handbook by Don Taylor	0-89586-402-9/$16.95
Mustang Weekend Projects 1964½–1967 by Jerry Heasley	1-55788-230-4/$17.00
Paint & Body Handbook (Revised Edition) by Don Taylor and Larry Hofer	1-55788-082-4/$16.95
Race Car Engineering & Mechanics by Paul Van Valkenburgh	1-55788-064-6/$16.95
Sheet Metal Handbook by Ron and Sue Fournier	0-89586-757-5/$16.95
Street Rodder's Handbook by Frank Oddo	0-89586-369-3/$16.95
Turbo Hydra-matic 350 by Ron Sessions	0-89586-051-1/$16.95
Turbochargers by Hugh MacInnes	0-89586-135-6/$16.95
Welder's Handbook (Revised Edition) by Richard Finch	1-55788-264-9/$16.95
Mustang 5.0 Projects by Mark Houlahan	1-55788-275-4/$16.95
The Car Builder's Handbook by Doug McCleary	1-55788-278-9/$16.95
Tri-Five Chevy Handbook by Jim Richardson	1-55788-285-1/$16.95
Performance Wheels and Tires by Mike Mavrigian	1-55788-286-X/$16.95

TO ORDER CALL: 1-800-788-6262, Refer to Ad #583b

HPBooks
A member of Penguin Putnam Inc.
375 Hudson Street
New York, NY 10014

*Prices subject to change